Lecture Notes in Physics

Edited by J. Ehlers, München, K. Hepp, Zürich, and
H. A. Weidenmüller, Heidelberg, and J. Zittartz, Köln
Managing Editor: W. Beiglböck, Heidelberg

53

D. J. Simms
N. M. J. Woodhouse

Lectures on Geometric Quantization

Springer-Verlag
Berlin · Heidelberg · New York 1976

Authors
Dr. D. J. Simms
School of Mathematics
Trinity College
Dublin 2/Ireland

Dr. N. M. J. Woodhouse
Mathematical Institute
University of Oxford
24–29 St. Giles
Oxford OX1 3LB/Great Britain

Library of Congress Cataloging in Publication Data

Simms, David John.
Lectures on geometric quantization.

(Lecture notes in physics ; 53)
1. Quantum theory. 2. Geometry, Differential.
I. Woodhouse, Nicholas Michael John, 1949- joint
author. II. Title. III. Series.
QC174.12.S53 530.1'2 76-26980

ISBN 3-540-07860-6 Springer-Verlag Berlin · Heidelberg · New York
ISBN 0-387-07860-6 Springer-Verlag New York · Heidelberg · Berlin

Printing and binding: Beltz Offsetdruck, Hemsbach/Bergstr.

Contents

1 Introduction....1

2 Symplectic Geometry and Hamiltonian Mechanics....5

3 The Lie Algebra $C^\infty(M)$: Poisson Brackets....14

4 Observables....18

5 Hermitian Line Bundles....23

6 Prequantization....38

7 Quantization....51

8 Invariance Groups....86

9 Examples....94

Appendix A....121

Appendix B....135

Appendix C....146

Acknowledgements

David Simms would like to thank F.A.E.Pirani and R.F.Streater for their hospitality at King's and Bedford Colleges. Nicholas Woodhouse would like to thank the SRC for their financial support.

1. Introduction

"Nature likes theories which are simple when stated in coordinate-free, geometric language."[1]

These notes are based on a series of ten lectures given by Professor David Simms in London, in the autumn of 1974, in which he outlined the "geometric quantization" programme[2] of B. Kostant and J-M. Souriau. The aim of the programme is to find a way of formulating the relationship between classical and quantum mechanics in a geometric language, as a relationship between symplectic manifolds (classical phase spaces) and Hilbert spaces (quantum phase spaces). Though it is unlikely that a completely general and intrinsic quantization construction will ever be found, work on this programme has lead to new insights into the connection between the concepts of symmetry in classical and in quantum mechanics and into the ambiguities involved in the physicist's concept of quantization.

Much work is still to be done, but already a coherent picture is emerging, at least for systems with a finite number of degrees of freedom (much less is known about field theories, and these will not be discussed here): the passage from classical to quantum mechanics depends on the introduction into the classical phase space of an additional geometric structure called a polarization (§7). In certain cases, for example, if there is a natural configuration space or if there is a prescribed symmetry group (such as the Poincaré group), the choice of polarization is more or less fixed. But, in general, the quantization construction is not unique: the quantum systems obtained using different polarizations are identical only in the semi-classical limit.

These notes begin with a very brief account of symplectic geometry and its role in Hamiltonian mechanics and continue with a more or less self-contained description of this quantization construction and its dependence on the choice of polarization. The notetaker (N.W.) has added some material to the original lectures (in particular, a collection of examples (§9) and three appendices). In addition, some minor notational changes have been made.

It is assumed that the reader is familiar with the elements of differential geometry and the exterior calculus. The appendices contain brief accounts of some topics which are needed for a complete understanding of the main text and which are perhaps unfamiliar to some applied mathematicians and physicists.

Notation: On a smooth manifold M:

1) $\Omega(M)$ denotes the complex exterior algebra

2) $\Omega^k(M)$ denotes the space of complex k-forms

3) $C^\infty(M)$ denotes the ring of smooth complex valued functions and $C^\infty_{\mathbb{R}}(M)$ the subset of real functions

4) $\mathcal{U}(M)$ denotes the space of complex vector fields.

The Lie derivative is denoted $\mathcal{L}$. In the exterior calculus, the sign conventions used here are fixed by: let $\alpha \in \Omega^p(M)$, $\beta \in \Omega^q(M)$ and $\xi \in \mathcal{U}(M)$. In local coordinates $\{x^a\}$:

$$\alpha = \alpha_{abc} \ldots dx^a \otimes dx^b \otimes \ldots$$

$$\beta = \beta_{abc} \ldots dx^a \otimes dx^b \otimes \ldots$$

$$\xi = \xi^a \frac{\partial}{\partial x^a}$$

(using the summation convention) and:

1) $\alpha \wedge \beta = \alpha_{[ab} \ldots \beta_{cd} \ldots{}_{]} \; dx^a \otimes dx^b \otimes \ldots \in \Omega^{p+q}(M)$

2) $d\alpha = \partial_{[a}\, \alpha_{bc} \ldots{}_{]} \; dx^a \wedge dx^b \wedge \ldots \quad \in \Omega^{p+1}(M)$

3) $\xi \lrcorner\, \alpha = p.\ \alpha(\xi, ., \ldots, .)$

$\qquad = p.\ \xi^a\, \alpha_{abc} \ldots dx^b \wedge dx^c \wedge \ldots \quad \in \Omega^{p-1}(M).$

Here square brackets denote antisymmetrization and $\partial_a = \dfrac{\partial}{\partial x^a}$.

The (complexified) tangent bundle to M is denoted TM ($TM^{\mathbb{C}}$), the (complexified) cotangent bundle, T*M ($T^*M^{\mathbb{C}}$), and the (complexified) tangent space at $m \in M$, $T_m M$ ($T_m M^{\mathbb{C}}$).

The transpose of a matrix a is denoted ${}^T a$.

2. Symplectic Geometry and Hamiltonian Mechanics

The geometric formulation of classical mechanics begins with the concept of a symplectic manifold, that is a pair (M,ω) of which

i) M is a smooth manifold

ii) ω is a real nondegenerate closed 2-form on M.

In local coordinates $\{x^a\}$, ω is given by:

$$\omega = \omega_{ab}\, dx^a \wedge dx^b \qquad 2.1$$

where

$$\partial_{[a}\omega_{bc]} = 0 \qquad \text{(since } \omega \text{ is closed)} \qquad 2.2$$

$$\det(\omega_{ab}) \neq 0 \qquad \text{(since } \omega \text{ is nondegenerate)} \qquad 2.3$$

As an example, consider the Cartesian space $M = \mathbb{R}^{2n}$ with the natural coordinates $\{q^1, q^2 \ldots q^n,\ p_1, p_2 \ldots p_n\}$ and the canonical symplectic 2-form

$$\omega = dp_a \wedge dq^a \qquad 2.4$$

In this case:

$$2\left[\omega_{ab}\right] = \begin{bmatrix} 0 & 1_n \\ -1_n & 0 \end{bmatrix} \qquad 2.5$$

where 1_n is the $n \times n$ unit matrix : ω is certainly closed and nondegenerate.

This example is basic in that every symplectic manifold looks like this locally; stated precisely:

Theorem[3] (Darboux): Every symplectic manifold (M,ω) admits an atlas of canonical coordinates.

That is, near each point $m \in M$, there are local coordinates $\{q^1, \ldots, q^n, p_1, \ldots, p_n\}$ (called canonical coordinates) in which ω takes the form:

$$\omega = dp_a \wedge dq^a \quad . \qquad 2.6$$

The theorem can be paraphrased by saying that if ω is thought of as a (skew symmetric) metric tensor then it follows from the closure of ω that M is essentially flat. Put another way, locally any symplectic manifold is completely determined by its dimension (which must be even since ω is non-degenerate).

The ideas of symplectic geometry arise in mechanics because the (momentum) phase space of a classical system carries a natural symplectic structure (which is closely related to the Poisson bracket concept). For example, let X be the configuration space of a holonomic, conservative, time independent classical dynamical system with finitely many degrees of freedom. It will be assumed that X is a C^∞ manifold. The velocity phase space of the system is then represented by the (real) tangent bundle of X, that is the set TX of pairs (x, ξ) where $x \in X$ and $\xi \in T_x X$ is a real tangent vector at x.

The dynamics and the transition to momentum phase space and the Hamiltonian formulation are determined by a Lagrangian, that is by a real function:

$$L : TX \to \mathbb{R}$$

on velocity phase space. Explicitly, the canonical momenta are given by the fibre derivative of L: for each $(x,\xi) \in TX$, the restriction of L to the fibre $T_x X$ is a function on the fibre, so its gradient

$$d(L|_{T_x X})$$

is a 1-form on the fibre and

$$d(L|_{T_x X})(x,\xi)$$

is an element of the dual space $T_x^* X$ to $T_x X$, that is, a covector at x. Now the cotangent bundle $T^* X$ is the set of all pairs (x, p) where $x \in X$ and $p \in T_x^* X$, so the map (called the <u>fibre derivative</u> of L) defined by:

$$FL: (x,\xi) \longmapsto (x, d(L|_{T_x X})(x,\xi))$$

is a map from velocity phase space to $T^* X$: in physical terms $T^* X$ is the

<u>momentum phase space</u> and

$$p = d(L|_{T_xX})(x,\xi)$$

is the <u>canonical momentum</u> corresponding to (x,ξ). In classical mechanics, FL is referred to as the <u>Legendre transformation</u>. It is usual to assume that FL is a diffeomorphism.

The function $h = A - L : T^*X \to \mathbb{R}$, where

$$A(x,p) = \xi \lrcorner\, p \;\; ; \;\; (x,\xi) = FL^{-1}(x,p),$$

is called the <u>Hamiltonian</u> of the system: it will be shown later how h generates the orbits of the system in (momentum) phase space.

Now if pr denotes the projection:

$$pr : TX \to X : (x,\xi) \longmapsto x$$

and if $\{x^a\}$ are local coordinates on X then each ξ can be written $\xi = v^a \dfrac{\partial}{\partial x^a}$, so that $q^a = x^a \circ pr$ and v^a can be taken as local coordinates[4] on TX. Then:

$$L(x,\xi) = L(q^a,v^a) \tag{2.7}$$

and

$$d(L|_{T_xX}) = \frac{\partial L}{\partial v^a}\, dv^a \tag{2.8}$$

If pr also denotes the projection:

$$pr : T^*X \to X : (x,p) \longmapsto x$$

then each p can be written $p = p_a\, dx^a$ and so $q^a = x^a \circ pr$ and p_a can be used as local coordinates on T^*X. With these (customary) choices, the Legendre map takes on the coordinate form:

$$FL : (q^a, v^a) \longmapsto (q^a, p_a) = (q^a\ ,\ \frac{\partial L}{\partial v^a})$$

which is familiar from elementary mechanics.

In this system, the natural symplectic structure referred to above is constructed from the canonical 1-form θ on T^*X. This may be defined in several equivalent ways:

1) If $\tau \in T_{(x,p)}(T^*X)$ is a tangent vector to T^*X at (x,p) then θ is defined at (x,p) by its value on τ :

$$\tau \lrcorner\, \theta_{(x,p)} = p\ (\ pr_*\ \tau\ ) \qquad 2.9$$

where pr_* is the induced map of tangent vectors defined by pr.

2) Equivalently, if $\alpha : X \to T^*X$ is any section of T^*X (that is, any 1-form on X) then θ is the unique 1-form on T^*X satisfying:

$$\alpha^*(\theta) = \alpha\quad . \qquad 2.10$$

3) Equivalently, in the local coordinates $\{q^a, p_a\}$ defined above:

$$\theta = p_a \, dq^a \tag{2.11}$$

The natural symplectic structure ω on T^*X is given by

$$\omega = d\theta \tag{2.12}$$

This 2-form is certainly closed; it is also non-degenerate, since, in the coordinates introduced above:

$$\omega = dp_a \wedge dq^a \tag{2.13}$$

(from which it also follows that these coordinates are canonical).

In the next section, it will be shown how this symplectic structure is related to the classical concept of the Poisson bracket. First, however, it is necessary to point out that (T^*X, ω) has a particularly simple structure since ω is not only closed but exact (since, $\omega = d\theta$) and that other symplectic manifolds for which this is <u>not</u> true also arise in classical mechanics (for instance, as the phase spaces of particles with internal degrees of symmetry, see §9). In particular, the concept of a Kähler manifold makes frequent appearances.

A simple example of a Kähler manifold is complex projective space $M = P^n\mathbb{C}$. This is defined as the set of rays in $\mathbb{C}^{n+1}$, that is as the quotient of $\mathbb{C}^{n+1} - \{0\}$ by the equivalence relation:

$$(z^1, \ldots z^{n+1}) \sim (w^1, \ldots w^{n+1})$$

whenever:

$$z^i = \lambda w^i \ ; \quad \lambda \in \mathbb{C}^* = \mathbb{C} - \{0\} \qquad 2.14$$

One can introduce local complex coordinates on M by:

$$Z^a = z^a/z^{n+1} \qquad a = 1,2 \ldots n \qquad 2.15$$

The domain of these is the whole of M less the hyperplane on which $z^{n+1} = 0$ (of course, by successively interchanging z^{n+1} and each z^j one defines coordinate patches covering this hyperplane also). The real and imaginary parts of the Z^a's can be used as real local coordinates on M: thus M is an n-dimensional complex manifold and a 2n dimensional real manifold.

There is a natural symplectic structure on M. In these local complex coordinates it is given by:

$$\omega = i \frac{\partial^2 f}{\partial Z^a \, \partial \overline{Z^b}} dZ^a \wedge d\overline{Z^b} \qquad 2.16$$

where:

$$f(Z^1, \ldots Z^n) = \ln(1 + \sum_{a=1}^{n} |Z^a|^2) \ . \qquad 2.17$$

A short calculation reveals that ω is closed and non-degenerate; however it is not exact (since M is compact[5]).

More generally, an almost Kähler manifold is defined to be a triple (M,J,ω) of which (M,ω) is a symplectic manifold and J is a tensor field of type $\binom{1}{1}$: that is, for each $m \in M$

$$J: T_mM \to T_mM$$

is a linear map. This tensor field must satisfy:

1) $J^2 = -1$ (a tensor field with this property is called an <u>almost complex structure</u>)

2) The bilinear form g defined by:

$$g(\xi,\zeta) = \omega(\xi,J\zeta); \quad \xi,\zeta \in T_mM \qquad 2.18$$

is a (real) positive definite Riemannian metric.

3) For each $m \in M$:

$$\omega(J\xi,J\zeta) = \omega(\xi,\zeta) \qquad \forall \ \xi,\zeta \in T_mM \qquad 2.19$$

In particular, complex projective space is an almost Kähler manifold with the almost complex structure defined by:

$$J\left(\frac{\partial}{\partial Z^a}\right) = i\,\frac{\partial}{\partial Z^a} \qquad J\left(\frac{\partial}{\partial \bar{Z}^a}\right) = -\,i\,\frac{\partial}{\partial \bar{Z}^a} \qquad 2.20$$

or, if $Z^a = x^a + iy^a$, by

$$J\left(\frac{\partial}{\partial x^a}\right) = \frac{\partial}{\partial y^a} \quad J\left(\frac{\partial}{\partial y^a}\right) = -\,\frac{\partial}{\partial x^a}\,. \qquad 2.21$$

If (M,J,ω) can be given the structure of a complex manifold in such a way that, in local complex coordinates, J is given as in eqn. 2.20 then it is called a <u>Kähler manifold</u>. It follows from a theorem[6] of Newlander and Nirenberg that this will be so if and only if the tensor field S (the <u>torsion</u> of J) vanishes, where

$$S(\xi,\zeta) = [\xi,\zeta] + J[J\xi,\zeta] + J[\xi,J\zeta] - [J\xi,J\zeta] \qquad 2.22$$

(ξ and ζ are vector fields on M).

Both the cotangent bundle and the Kähler manifold are exceptional examples of symplectic manifolds in that they both have naturally defined polarizations: this will be explained in §7.

3. The Lie Algebra $C^\infty(M)$: Poisson Brackets

On any symplectic manifold (M,ω), the symplectic form defines an isomorphism:

$$T_m M \to T_m^* M \quad : \quad \xi \longmapsto -\xi \lrcorner \omega \quad = -2\omega(\xi,\cdot)$$

between the tangent and cotangent spaces at each point $m \in M$. There is thus a natural identification between $\mathcal{U}(M)$ (the space of complex vector fields on M) and $\Omega^1(M)$ (the space of complex 1-forms). Elements of $\mathcal{U}(M)$ corresponding to exact 1-forms are called globally Hamiltonian vector fields, and those corresponding to closed 1-forms, locally Hamiltonian vector fields.

This relationship can be expressed diagramatically:

$$\begin{array}{ccccccccccc}
 & & & & & & \mathrm{Im}(d) & \hookrightarrow & \mathrm{Ker}(d) & \hookrightarrow & \Omega^1(M) & \xrightarrow{\ d\ } & \Omega^2(M) \\
 & & & & & \overset{d}{\nearrow} & \updownarrow & & \updownarrow & & \updownarrow & & \\
0 & \to & \mathbb{C} & \hookrightarrow & C^\infty(M) & \longrightarrow & A(M) & \hookrightarrow & A_0(M) & \hookrightarrow & \mathcal{U}(M) & &
\end{array}$$

Here $A(M)$ is the set of globally Hamiltonian vector fields and $A_0(M)$ is the set of locally Hamiltonian vector fields; $\mathrm{Im}(d) \subset \Omega^1(M)$ is the image of d, that is the set of exact 1-forms, and Ker(d) the set of closed 1-forms. The horizontal arrow $C^\infty(M) \to A(M)$ is the map $\phi \longmapsto \xi_\phi$ where $\phi \in C^\infty(M)$ and ξ_ϕ is the globally Hamiltonian vector field given by:

$$\xi_\phi \lrcorner\, \omega + d\phi = 0 \qquad 3.1$$

If ϕ is the (real) Hamiltonian function of a system with phase space (M,ω) then the orbits[7] of the system in M are the integral curves of ξ_ϕ : it is in this way that the Hamiltonian generates the dynamics of the system.

With each pair of functions $\phi,\psi \in C^\infty(M)$, there is associated a third function

$$[\phi,\psi] = \xi_\phi(\psi) \qquad 3.2$$

called the Poisson bracket[8] of ϕ and ψ: the Poisson bracket gives $C^\infty(M)$ the structure of a Lie algebra over $\mathbb{C}$. The globally Hamiltonian vector fields also form a (complex) Lie algebra (with the Lie bracket as skew product) and the map $\phi \longmapsto \xi_\phi$ is a Lie algebra homomorphism (that is, it is linear over $\mathbb{C}$ and preserves brackets).

To prove these statements, first note that eqn. 3.2 can be re-written:

$$[\phi,\psi] = \xi_\phi(\psi) = - \xi_\phi \lrcorner\, (\xi_\psi \lrcorner\, \omega) = 2\omega(\xi_\phi,\xi_\psi) \qquad 3.3$$

so that the Poisson bracket is certainly skew symmetric and bilinear (over $\mathbb{C}$). In local canonical coordinates $\{q^a,p_a\}$, ξ_ϕ is given by:

$$\xi_\phi = \frac{\partial\phi}{\partial p_a}\frac{\partial}{\partial q^a} - \frac{\partial\phi}{\partial q^a}\frac{\partial}{\partial p_a} \qquad 3.4$$

and the Poisson bracket assumes its familiar form:

$$[\phi,\psi] = \frac{\partial\phi}{\partial p_a}\frac{\partial\psi}{\partial q^a} - \frac{\partial\phi}{\partial q^a}\frac{\partial\psi}{\partial p_a} \qquad 3.5$$

The other properties of the Poisson bracket are most conveniently established using the relations:

$$\left.\begin{aligned} \mathcal{L}_\xi &= \xi \lrcorner\, d + d\, \xi \lrcorner & 3.6 \\ [\xi,\eta] &= \mathcal{L}_\xi\, \eta \lrcorner - \eta \lrcorner\, \mathcal{L}_\xi & 3.7 \end{aligned}\right\} \quad \eta,\xi \in \mathcal{U}(M)$$

between[9] $\mathcal{L}_\xi$, d and $\xi \lrcorner$, regarded as operators acting to the right on the exterior algebra $\Omega(M)$.

Two lemmas are needed:

3.1 <u>Lemma</u>: If $\eta \in \mathcal{U}(M)$ then $\eta \in A_o(M)$ if, and only if, $\mathcal{L}_\eta \omega = 0$.

<u>Proof</u>:

$$\eta \in A_o(M) \iff d(\eta \lrcorner\, \omega) = 0$$

$$\iff \mathcal{L}_\eta\omega - \eta \lrcorner\, d\omega = 0 \qquad \text{(using 3.6)}$$

$$\iff \mathcal{L}_\eta\omega = 0 \qquad \text{(since } \omega \text{ is closed).}$$

3.2 <u>Lemma</u>: If $\zeta,\eta \in A_o(M)$ then $[\zeta,\eta] = 2\xi_{\omega(\zeta,\eta)} \in A(M)$.

<u>Proof</u>:

$$[\zeta,\eta] \lrcorner\, \omega = \mathcal{L}_\zeta(\eta \lrcorner\, \omega) - \eta \lrcorner\, \mathcal{L}_\zeta\, \omega \qquad \text{(using 3.7)}$$

$$= d(\zeta \lrcorner\, (\eta \lrcorner\, \omega)) + \zeta \lrcorner\, (d(\eta \lrcorner\, \omega)) \qquad \text{(using 3.6 and lemma 3.1)}$$

$$= 2\, d(\omega(\eta,\zeta))$$

$$= 2(\xi_{\omega(\zeta,\eta)}) \lrcorner\, \omega \qquad \square.$$

It follows from lemma 3.2 that the commutator of two Hamiltonian vector fields is globally Hamiltonian and that:

$$[\xi_\phi,\xi_\psi] = 2\xi_{\omega(\xi_\phi,\xi_\psi)} = \xi_{[\phi,\psi]} \qquad 3.8$$

that is, the map $\phi \longmapsto \xi_\phi$ preserves brackets. Thus all that remains to be shown is that the Poisson bracket satisfies the Jacobi identity. In fact, this is a direct consequence of the closure of ω:

$$0 = d\omega(\xi_\phi,\xi_\psi,\xi_\chi) \quad ; \qquad \phi,\psi,\chi \in C^\infty(M)$$

$$= \sum(\xi_\phi(\omega(\xi_\psi,\xi_\chi)) - \omega([\xi_\psi,\xi_\chi],\xi_\phi))$$

$$= \frac{1}{2} \quad \sum([\phi,[\psi,\chi]] - [[\psi,\chi],\phi])$$

$$= \sum[\phi, [\psi,\chi]] \qquad 3.9$$

where $\sum$ indicates cyclic summation over ϕ,ψ and χ.

4. Observables

In classical mechanics, an observable plays two roles. In the first place, it is a measurable quantity, represented by a smooth function on phase space; in the second place, at least locally, it is the generator of a one parameter family of canonical transformations. For example, in Newtonian mechanics, the fundamental observables, energy, momentum and angular momentum, act in phase space as the generators of time-translations, space-translations and rotations.

This duality emerges very clearly from the geometrical machinery developed in the last section. For suppose that (M,ω) is the phase space of some classical system. Then, associated with each smooth real function $\phi \in C^{\infty}_{\mathbb{R}}(M)$, there is a real vector field[10] $\xi_{\phi} \in \mathcal{U}(M)$ which satisfies (lemma 3.1):

$$\mathcal{L}_{\xi_{\phi}} \omega = 0 \qquad 4.1$$

This is precisely the condition that the local diffeomorphisms of M generated by ξ_{ϕ} should be canonical, that is that they should preserve ω. For certain observables (those for which ξ_{ϕ} is complete) this works globally: they generate canonical transformations of the whole of M.

Going in the other direction, however, that is deciding which 'measurable quantity' is associated with a given one parameter group of canonical transformations, is a more subtle problem; a detailed treatment will be postponed until §8. For the moment, I shall just illustrate the nature of the problem (which is essentially the same in quantum mechanics[11]) with a familiar example.

Suppose that the rotation group G = SO(3) acts on M as a group of canonical transformations. Each element of the Lie algebra $\mathcal{G}$ of G generates a one parameter subgroup of G; hence, by lemma 3.1, each $X \in \mathcal{G}$ gives rise to a locally Hamiltonian vector field $\xi_X \in A_o(M)$. Explicitly,

$$(\xi_X(\phi))\ (m) = \frac{d}{dt}\ \{\phi((\exp(-\ t.X)).m)\}\Big|_{t=o} \quad ; m \in M, \quad \phi \in C^\infty(M) \qquad 4.2$$

From the definition of the Lie bracket in $\mathcal{G}$, one has:

$$[\xi_X, \xi_Y] = \xi_{[X,Y]} \qquad X,Y \in \mathcal{G} \qquad 4.3$$

so that the map σ defined by:

$$\sigma : \mathcal{G} \to A_o(M) : X \mapsto \xi_X$$

is a homomorphism of Lie algebras.

Now the rotation group SO(3) has the special property that its Lie algebra satisfies[12]:

$$\mathcal{G} = [\mathcal{G}, \mathcal{G}] \qquad ; \qquad 4.4$$

In fact, each $Z \in \mathcal{G}$ is of the form:

$$Z = [X,Y] \quad ; \quad X,Y \in G \qquad 4.5$$

(For example, the generator of a "rotation about the z-axis" is the commutator of the generators of "rotations about the x- and y-axis".) But, from lemma 3.2, if $Z \in G$ is of the form $Z = [X,Y]$, then the vector field

$$\sigma(Z) = \xi_{[X,Y]} = [\xi_X, \xi_Y] \qquad 4.6$$

is <u>globally</u> Hamiltonian. Thus, in this case, σ maps G into the algebra $A(M)$ of globally Hamiltonian vector fields.

The problem of associating a measurable quantity with each one parameter subgroup of SO(3) is thus reduced to the problem of choosing a suitable map

$$\lambda : G \to C^{\infty}_{\mathbb{R}}(M)$$

which makes the diagram:

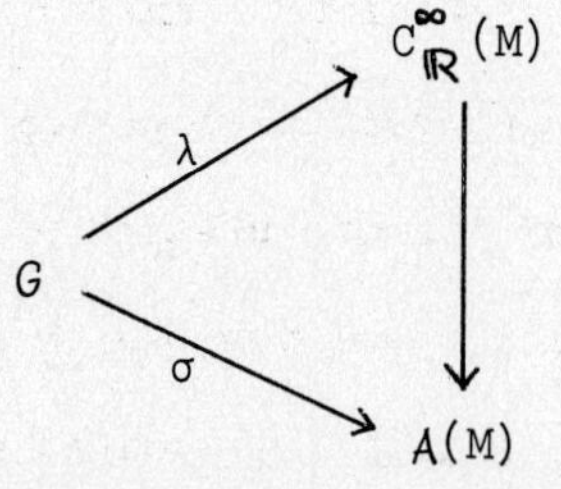

commute. Fortunately, in this case there is a preferred map. It follows from another special property of SO(3), namely that the first and second real cohomology groups of its Lie algebra are trivial (see appendix C), that there is a unique choice for λ which makes this diagram into a commutative diagram of Lie algebra homomorphisms. With this choice, each $X \in G$ corresponds to a smooth function $\phi_X = \lambda(X) \in C^{\infty}_{\mathbb{R}}(M)$ such that:

1) $$\xi_X = \xi_{X_\phi} \tag{4.7}$$

2) $$\phi_{[X,Y]} = [\phi_X, \phi_Y]\,; \ \forall\, X,Y \in G \tag{4.8}$$

(The situation is not always so clear cut: for example, in the case of the Galilei group one must first go to a central extension of the Lie algebra before constructing λ. More about this below).

In a purely classical context, there is no overwhelming reason for insisting that λ should be a Lie algebra homomorphism, that is that condition (2) (eqn. 4.8) should hold. However, this condition does play a central role in the quantization procedure. Essentially, it is equivalent to the requirement of naturality, that is that a symmetry group in a classical system should also be a symmetry group in the underlying quantum system. This will be clarified later.

The first stage of the quantization programme is to construct a Hilbert space representation of the Lie algebra of classical observables; that is, to construct a Hilbert space H on which each classical observable (that is, each element of $C^{\infty}_{\mathbb{R}}(M)$) is represented as a Hermitian operator in such a way that the Poisson bracket of two classical observables is represented by the commutator of the corresponding quantum operators. In its simplest form, the idea which will be developed below is to construct H from the space of complex valued functions on phase space. Indeed, this is precisely what is done when the phase space is the cotangent bundle of some configuration space. However, if more complicated systems are to be included (such as particles with internal degrees of freedom) it is necessary to modify this a little: H is constructed not from the complex valued functions but from the sections of a certain line bundle over phase space. The additional formalism needed for this modification is introduced in the next section; the construction itself will be described in §6.

As a byproduct, this approach yields a synthesis of the classical and quantum concepts of observable. In the construction, each observable appears as a vector field on the line bundle and from this representation one can derive with equal simplicity its three roles as a function on the classical phase space, as a generator of a family of canonical transformations and as an operator on the quantum Hilbert space.

5. Hermitian Line Bundles

Informally, a line bundle over a smooth manifold is a 'twisted' Cartesian product of the manifold with the complex numbers. In the more precise language of the general theory of bundles (which is outlined in appendix B) a line bundle is a vector bundle with a one dimensional complex vector space as fibre.

Any line bundle has the property that it is equivalent, if its zero section is deleted, to its associated principal bundle. Because of this, it is possible to state the defining properties in a relatively simple form and to bypass, for the moment, some of the technicalities of the general theory. Explicitly, a line bundle over a smooth manifold M is a triple (L, π, M) where:

(1) L is a smooth manifold and π is a smooth map of L onto M

(2) For each $m \in M$, $L_m = \pi^{-1}(m)$ has the structure of a one dimensional complex vector space (L_m is called the <u>fibre</u> over m)

(3) There is an open cover $\{U_j \mid j \in \Lambda\}$ of M (indexed by some set Λ) and a collection of C^∞ maps $s_j : U_j \to L$ such that:

   (a) For each $j \in \Lambda$, $\pi \circ s_j = 1_{U_j}$ (the identity map on U_j).

   (b) For each $j \in \Lambda$, the map $\psi_j : U_j \times \mathbb{C} \to \pi^{-1}(U_j) : (m,z) \longmapsto z.s_j(m)$ is a diffeomorphism (the multiplication is scalar multiplication in the vector space L_m).

The collection $\{U_j, s_j\}$ is called a <u>local system</u> for L.

A smooth map $s : U \to L$ from some open subset of M into L which satisfies

$$\pi \circ s = 1_U \qquad 5.1$$

is called a <u>local section</u> of L (or simply a <u>section</u> if U = M). Thus a local system is a cover of M by non-vanishing local sections.

The set $\Gamma(L)$ of all sections of L forms a $C^\infty(M)$-module, with the multiplication law:

$$(\phi s)(m) = \phi(m) \, . \, s(m); \quad \phi \in C^\infty(m), \quad s \in \Gamma(L) \qquad 5.2$$

Here again, the multiplication on the right hand side is scalar multiplication in L_m.

The simplest example of a line bundle is the trivial or product bundle with $L = M \times \mathbb{C}$ and $\pi : L \to M$ the projection onto the first factor. As a local system for L, one can take the set $\{(M, s_o)\}$ where:

$$s_o : M \to L : \quad m \longmapsto (m, 1)$$

is the unit section of L; $\Gamma(L)$ is then isomorphic with $C^\infty(M)$ since any other section s is uniquely of the form:

$$s = \phi \,.\, s_o \qquad 5.3$$

for some $\phi \in C^\infty(M)$. In the general case, it is helpful to think of a section as a generalized complex valued function.

The line bundles used in geometric quantization have two additional structures:

1) A Hermitian metric : on each fibre there is a Hilbert space metric $(\cdot\,,\cdot)$ with the property that, for any $s, t \in \Gamma(L)$, the function (s,t) defined by:

$$(s,t) : M \to \mathbb{C} : m \longmapsto (s(m), t(m))$$

is smooth.

2) A connection : there is a map ∇ which assigns to each vector field $\xi \in \mathcal{U}(M)$ an endomorphism $\nabla_\xi : \Gamma(L) \to \Gamma(L)$ satisfying:

(a) $\nabla_{\xi+\eta} s = \nabla_\xi s + \nabla_\eta s$ 5.4

(b) $\nabla_{\phi.\xi} s = \phi\, \nabla_\xi s$ 5.5

(c) $\nabla_\xi(\phi.s) = (\xi\phi).s + \phi.\nabla_\xi s$ 5.6

for each $s \in \Gamma(L)$, $\xi, \eta \in \mathcal{U}(M)$ and $\phi \in C^\infty(M)$. (The map $\nabla_\xi : \Gamma(L) \to \Gamma(L)$ is called the covariant derivative along ξ ; (a) and (b) are the usual requirements of linearity and (c) is the Leibnitz rule.)

These two structures are required to be compatible in the sense that, for each real $\xi \in \mathcal{U}(M)$

$$\xi(s,t) = (s, \nabla_\xi t) + (\nabla_\xi s,t) \qquad \forall\ s,t \in \Gamma(L) \tag{5.7}$$

A line bundle with a connection ∇ and a compatible Hermitian metric is called a <u>Hermitian line bundle with connection</u>.

For example, the trivial bundle has a natural Hermitian metric defined by:

$$((m,z_1), (m,z_2)) = z_1\bar{z}_2 \ ; \qquad (m,z_i) \in M \times \mathbb{C} \tag{5.8}$$

One may also construct a connection on this bundle by choosing any $\alpha \in \Omega^1(M)$ and putting:

$$\nabla_\xi s = (\xi\phi + 2\pi i(\xi \lrcorner \alpha) . \phi) s_o \tag{5.9}$$

where $s \in \Gamma(L)$ and $\phi \in C^\infty(M)$ is defined by:

$$s = \phi . s_o \tag{5.10}$$

This will be compatible with the natural Hermitian metric if, and only if, α is real.

This example is basic in the sense that, locally, any connection on a line bundle L has this form. To see this, choose a local system

$\{(U_j, s_j)\}$ for L and, for each j, consider the map:

$$\mathcal{U}(M) \to C^\infty(U_j) : \xi \longmapsto \frac{1}{2\pi i} \cdot s_j^{-1} \nabla_\xi s_j$$

which is linear in ξ over $C^\infty(M)$ and so defines a 1-form $\alpha_j \in \Omega^1(U_j)$. It then follows from eqn. 5.6 that

$$(\nabla_\xi s)|_{U_j} = (\xi \phi_j + 2\pi i \cdot (\xi \lrcorner \alpha_j) \cdot \phi_j) s_j \qquad 5.12$$

where $s \in \Gamma(L)$ and $\phi_j \in C^\infty(U_j)$ is defined by:

$$s = \phi_j \cdot s_j \quad \text{in} \quad U_j \quad . \qquad 5.13$$

The 1-form α_j can be thought of as a 'Christoffel symbol' for the connection.

On each non-empty intersection $U_j \cap U_k$, one has $s_j = c_{jk} \cdot s_k$ for some function $c_{jk} \in C^\infty(U_j \cap U_k)$, so that the corresponding 'Christoffel symbols' α_j and α_k are related by:

$$\alpha_j = \alpha_k + \frac{1}{2\pi i} \cdot \frac{d\, c_{jk}}{c_{jk}} \quad \text{on} \quad U_j \cap U_k \qquad 5.14$$

(The functions $\{c_{jk} \mid j,k \in \Lambda\}$ are called the transition functions of the local system).

Any collection of 1-forms $\alpha_j \in \Omega^1(U_j)$ which are related as in eqn. 5.14 will define a connection on L. This can be used to give an alternative and, in some cases, more convenient characterization of a connection as a 1-form not on M, but on the space L^* obtained by deleting the origin from each fibre of L. For suppose there is given a connection ∇, and hence a collection $\alpha_j \in \Omega^1(U_j)$ of 1-forms related as in eqn. 5.14. For each j, define $\beta_j \in \Omega^1(U_j \times \mathbb{C}^*)$ by:

$$\beta_j = pr_1^*(\alpha_j) + \frac{1}{2\pi i} \cdot \frac{dz}{z} \qquad 5.15$$

where $\mathbb{C}^* = \mathbb{C} - \{0\}$ and $pr_1 : U_j \times \mathbb{C}^* \to U_j$ is the projection onto the first factor. Under the diffeomorphism $\psi_j : U_j \times \mathbb{C} \to \pi^{-1}(U_j)$, β_j is mapped onto a 1-form $(\psi_j^{-1})^*(\beta_j)$ on $\pi^{-1}(U_j) \cap L^*$. It follows from eqn. 5.14, that, for each j,k, the 1-forms $(\psi_j^{-1})^*(\beta_j)$ and $(\psi_k^{-1})^*(\beta_k)$ are equal on $\pi^{-1}(U_j) \cap \pi^{-1}(U_k) \cap L^*$. Thus there is a well defined 1-form $\alpha \in \Omega^1(L^*)$ satisfying:

$$\psi_j^* \alpha = \beta_j = pr_1^*(\alpha_j) + \frac{1}{2\pi i} \cdot \frac{dz}{z} \in \Omega^1(U_j \times \mathbb{C}^*)$$

for each j. Furthermore, the α_j's, and hence the connection, are completely determined by α. In fact, the connection ∇ is given directly in terms of α by:

$$\nabla_\xi s = 2\pi i(\xi \lrcorner s^* \alpha) . s \quad ; \quad s \in \Gamma(L), \ \xi \in \mathcal{U}(M) \qquad 5.16$$

(at points where s = 0 this is only valid in the limit) since, on U_j:

$$s^* \alpha = \alpha_j + \frac{1}{2\pi i} \frac{d\phi}{\phi} \qquad 5.17$$

where:

$$s = \phi \, . \, s_j \, . \qquad 5.18$$

By its construction, α has the two properties:

(1) It is invariant under the action of $\mathbb{C}^*$ (each non-zero complex number z defines a diffeomorphism of L^* by scalar multiplication in the fibres. It follows from the expression for β_j (eqn. 5.15) that this diffeomorphism leaves α invariant.)

(2) For each $m \in M$, the pull-back of α under any non-singular linear map $\mathbb{C}^* \to L_m^* \subset L^*$ is $\frac{1}{2\pi i} \cdot \frac{dz}{z}$ (L_m^* is L_m with the origin deleted).

By reversing the argument, it is not hard to see that any 1-form $\alpha \in \Omega^1(L^*)$ with these two properties will define a connection via eqn. 5.16.

The compatibility condition (eqn. 5.7) can now be put in a more concise form. A Hermitian metric $(\cdot\ ,\ \cdot)$ on L is completely determined by the function:

$$H : L \to \mathbb{R} \quad : \ell \longmapsto (\ell, \ell) \, . \qquad 5.19$$

If ∇ is a connection on (L, π, M) then ∇ and $(\cdot\ ,\ \cdot)$ will be compatible if, and only if, H and the connection form α of ∇ are related by:

$$\frac{dH}{H} = 2\pi i(\alpha - \bar{\alpha}) \qquad 5.20$$

To see this, let $s: U \subset M \rightarrow L$ be any non-vanishing local section and let ξ be a real vector field on U. Then:

$$\xi(s,s) = \xi(H \circ s) = ((s_*\xi) \lrcorner\, dH) \circ s \qquad 5.21$$

and, from eqn. 5.16

$$(\nabla_\xi s,s) + (s,\nabla_\xi s) = 2\pi i(\xi \lrcorner\, (s^*\alpha - s^*\bar{\alpha})).\,(s,s)$$

$$= (2\pi i\, H\,.\,(s_*\xi) \lrcorner\, (\alpha - \bar{\alpha})) \circ s \qquad 5.22$$

whence it is clear that:

$$\xi(s,s) = (\nabla_\xi s,s) + (s,\nabla_\xi s) \qquad 5.23$$

for every choice of ξ and s if, and only if:

$$\frac{dH}{H} = 2\pi i\,(\alpha - \bar{\alpha}). \qquad 5.24$$

Suppose now that there is given a line bundle (L,π,M) and a connection ∇. In general, there will exist vector fields ξ, $\eta \in \mathcal{U}(M)$ for which the operators ∇_ξ and ∇_η do not commute and the connection will have curvature. Formally, given two vector fields $\xi, \eta \in \mathcal{U}(M)$, one defines the operator curv(L,∇) (ξ,η) by:

$$\mathrm{curv}(L,\nabla)\ (\xi,\eta)\ (s) = \frac{1}{2\pi i}\left([\nabla_\xi,\nabla_\eta] - \nabla_{[\xi,\eta]}\right) s \qquad 5.25$$

where $s \in \Gamma(L)$. The right hand side of this equation is skew symmetric in ξ and η and linear over $C^\infty(M)$ in ξ,η and s. Thus the left hand side must be of the form:

$$\mathrm{curv}(L,\nabla)\ (\xi,\eta)\ (s) = \Omega(\xi,\eta)\ .\ s \qquad 5.26$$

for some 2-form $\Omega \in \Omega^2(M)$; Ω is called the curvature 2-form of the connection.

It follows from eqn. 5.15 that:

$$\pi^*(\Omega) = d\alpha \qquad 5.27$$

and that, in each U_j:

$$\Omega|_{U_j} = d\alpha_j \qquad 5.28$$

so that Ω is closed.

However, the crucial result is that, for any connection ∇, the curvature 2-form $\Omega = \mathrm{curv}(L,\nabla)$ is always integral, that is, the result of integrating it over any closed two dimensional contour in

M is an integer. A somewhat technical proof of this result, which is called the integrality condition, is given in appendix A. However, it is possible to gain some geometrical insight into its meaning by introducing the idea of parallel transport.

A smooth curve Γ: $[a,b] \to L^*$ is said to be parallel (or <u>horizontal</u>) if its tangent vector Ξ satisfies:

$$\Xi \lrcorner \alpha = 0 \qquad 5.29$$

Alternatively, Γ is parallel if there exists a section s: $M \to L$ and a vector field $\xi \in \mathcal{U}(M)$ such that

1) $\xi = \pi_*(\Xi)$ on $\pi(\Gamma(|a,b|))$ 5.30

2) $\Gamma([a,b]) \subset s(M)$ 5.31

3) $\nabla_\xi s = 0$ on $\pi(\Gamma([a,b]))$ 5.32

If γ: $[a,b] \to M$ is any smooth curve and if $\ell_o \in L^*_{\gamma(a)}$ then there is a unique parallel curve Γ : $[a,b] \to L^*$ through ℓ_o satisfying :

$$\pi \circ \Gamma = \gamma \qquad 5.33$$

This is certainly true if $\gamma([a,b])$ is contained in some U_j since if

ξ is the tangent to γ and $\ell_o = z_o \cdot s_j(a)$ then Γ is given explicitly by:

$$\Gamma : t \longmapsto \psi_j(\gamma(t), z(t))$$

where:

$$z: [a,b] \to \mathbb{C}^* \; : \; t \longmapsto z(t)$$

is the unique solution of the differential equation:

$$\frac{\dot{z}(t)}{z(t)} = -2\pi i \cdot \xi(t) \lrcorner \alpha_j$$

with the initial condition:

$$z(o) = z_o \, .$$

It is also true globally since $\gamma([a,b])$ can always be covered by a finite subset of $\{U_j\}$.

If $\Gamma: [a,b] \to L^*$ is a parallel curve covering $\gamma : [a,b] \to M$ (that is, satisfying eqn. 5.33) then the point $\Gamma(b)$ is said to be reached from $\Gamma(a)$ by <u>parallel transport</u> along γ.

Now suppose that $\gamma : [a,b] \to M$ is a closed curve (so that $\gamma(a) = \gamma(b)$) spanned by a 2-surface W which is diffeomorphic with $\mathbb{R}^2$. Under parallel propagation around γ, each point $\ell \in L^*_{\gamma(a)}$ is mapped onto a second point $P(\ell) \in L^*_{\gamma(a)}$. The map P is linear in ℓ, so that there is a non zero complex number z_γ such that:

$$P(\ell) = z_\gamma . \ell \quad \forall \ \ell \in L^*_{\gamma(a)} \quad .$$

The trick is to express z_γ as a surface integral over W; first a lemma:

<u>Lemma</u>: Let (L,π,M) be a line bundle with connection ∇. If the open set $U \subset M$ is smoothly contractible to a point then there exists a nowhere vanishing local section $s: U \to L$.

<u>Proof</u>: That U is smoothly contractible means that there is a point $m_o \in U$ and a smooth map $F: U \times [0,1] \to U$ such that, for every $m \in U$:

1) $F(m,1) = m$

2) $F(m,0) = m_o$.

To construct $s: U \to L$, choose an arbitrary point $\ell_o \in L^*_{m_o}$ and, for each $m \in U$, define $s(m)$ by parallely propagating ℓ_o along the curve

$$t \longmapsto F(m,t) \qquad \square \, .$$

Now the surface W is smoothly contractible so it must lie in an open neighbourhood U which is also smoothly contractible. Let

$s: U \to L^*$ be a non-vanishing local section and let $\Gamma: [a,b] \to L^*$ be the unique parallel curve through $s(\gamma(a))$ which satisfies:

$$\pi \circ \Gamma = \gamma \quad .$$

Since s is non-vanishing, Γ is given by

$$\Gamma(t) = \psi(t) \,.\, s(\gamma(t)); \quad t \in [a,b]$$

for some function $\psi : [a,b] \to \mathbb{C}^*$. The condition that Γ be parallel then translates into the form

$$\dot{\phi}(t) = - 2\pi i(\xi(t) \lrcorner\ s^* \alpha).\ \phi(t)$$

where ξ is the tangent vector to γ. Hence:

$$z_\gamma = \phi(b)/\phi(a) = \exp(- 2\pi i \oint (\xi \lrcorner\ s^* \alpha)\, dt)$$

$$= \exp(- 2\pi i \oint s^* \alpha)$$

the integral being taken around γ. By Stokes' theorem, this contour integral can be written as a surface integral over W. The result is:

$$z_\gamma = \exp\,(- 2\pi i \int_W d(s^* \alpha)) = \exp\,(- 2\pi i \int_W \Omega)$$

where the orientation of W is chosen to be compatible with that of γ.

It is now possible to understand the geometrical origin of the integrality condition. Suppose, for example, that S is a closed 2-surface in M diffeomorphic with the sphere S^2. Choose a smooth closed curve $\gamma : [a,b] \to M$ on S which divides S into the union of two 2-surfaces W_1 and W_2, each diffeomorphic with $\mathbb{R}^2$, (together with their common boundary $\gamma([a,b])$). By the argument above, the complex number z_γ which determines the result of parallel propagation around γ is given by the two expressions:

$$1) \qquad z_\gamma = \exp(-\,2\pi i \int_{W_1} \Omega)$$

$$2) \qquad z_\gamma = \exp(-\,2\pi i \int_{W_2} \Omega)$$

Taking into account the orientations of W_1 and W_2, it follows that:

$$\exp(-\,2\pi i \int_S \Omega) = 1$$

so that $\int_S \Omega$ must be an integer.

Geometric quantization begins with the converse of this result, which is the content of Weil's theorem.

Theorem[13] (Weil). If Ω is a closed real integral 2-form on a manifold M then there exists a line bundle (L,π,M) with a Hermitian metric $(\,.\,,\,.\,)$ and a compatible connection ∇ such that $\mathrm{curv}(L,\nabla) = \Omega$.

A proof of this is also given in appendix A together with a prescription for determining the number of essentially different line bundles with these properties in terms of the topological data of M. In particular, it is shown that if M is simply connected then ∇ and $(. , .)$ are unique up to equivalence. The concept of equivalence used here is the following: two line bundles (L_i, π_i, M_i) with connections ∇_i and metrics, $(. , .)_i$, defined by functions $H_i : L_i \to \mathbb{R}$, $(i = 1,2)$ are <u>equivalent</u> if there exists a diffeomorphism

$$\tau : L_1 \to L_2$$

which:

1) commutes with projection:

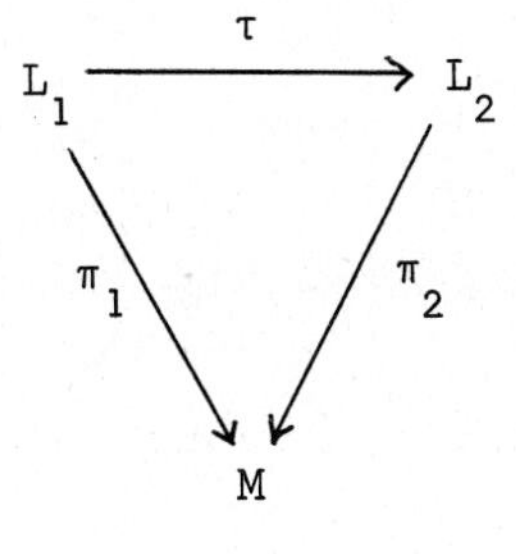

2) Restricts to a linear isomorphism

$$\tau_m : (L_1)_m \to (L_2)_m$$

for each $m \in M$.

3) Satisfies $\tau^*(\alpha_2) = \alpha_1$ and $H_2 \circ \tau = H_1$.

6. Prequantization

A symplectic manifold is said to be quantizable if its symplectic form is integral. The justification for the use of this suggestive term is this: if (M,ω) is quantizable then, by Weil's theorem, it is possible to find a Hermitian line bundle (L,π,M) with connection ∇ and curvature 2-form equal to ω. In this section, it will be shown that there is then a natural way of constructing a Hilbert space $\mathcal{H}$ from the space $\Gamma(L)$ of sections of L and that there is a natural representation of the Lie algebra $C^{\infty}_{\mathbb{R}}(M)$ by Hermitian operators on $\mathcal{H}$. Thus if a classical system has a quantizable phase space, this procedure (it is called prequantization) can be used to construct the Hilbert space and observables of a corresponding quantum system. Unfortunately, this is not the whole story. There is a difficulty concerned with reducibility of the representation: this difficulty, and a way of circumventing it, will be dealt with later.

In detail, prequantization works like this. Suppose that (M,ω) is a quantizable symplectic manifold and that (L,π,M) is a Hermitian line bundle over M with connection ∇ and curvature 2-form equal to ω. The set $\Gamma(L)$ of smooth sections of L forms a vector space over $\mathbb{C}$ under the operations:

$$\left.\begin{array}{ll}(1) & (s_1 + s_2)(m) = s_1(m) + s_2(m) \\ (2) & (z.s_1)(m) = z.s_1(m)\end{array}\right\} \quad \begin{array}{ll} s_1, s_2 \in \Gamma(L)\ , \quad m \in M & 6.1 \\ z \in \mathbb{C}\ . & 6.2\end{array}$$

Now there is a natural volume element on M given in terms of the symplectic form by[14]:

$$\omega^n = \omega \wedge \ldots \wedge \omega \qquad 6.3$$

This is used to define the inner product on $\Gamma(L)$: explicitly, if $s,t \in \Gamma(L)$ then $\langle s,t\rangle$ is the (not necessarily finite) complex number:

$$\langle s,t\rangle = \int_M (s(m), t(m))\, \omega^n. \qquad 6.4$$

The subspace of $\Gamma(L)$ of sections s for which $\langle s,s\rangle$ is finite forms a pre-Hilbert space H_o; the Hilbert space of prequantization is the completion H of H_o.

The first step in the construction of the quantum operators on H is to replace the classical Lie algebra of observables $C^\infty_{\mathbb{R}}(M)$ by a natural isomorph, namely the set $\varepsilon(L,\nabla)$ of all real vector fields η on L^* which have the properties:

(1) They are invariant under the action of $\mathbb{C}^*$

(2) They satisfy:

$$\mathcal{L}_\eta \alpha = 0 \text{ and } \eta H = 0 \qquad 6.5$$

where α is the connection form and $H: L^* \to \mathbb{R}$ is the function:

$$H : \ell \longmapsto (\ell,\ell).$$

(This set forms a Lie algebra under the Lie bracket operation on vector fields.)

The isomorphism between $C^{\infty}_{\mathbb{R}}(M)$ and $\varepsilon(L,\nabla)$ is constructed by mapping each classical observable $\phi \in C^{\infty}_{\mathbb{R}}(M)$ onto the unique real vector field $\eta_{\phi} \in \mathcal{U}(L^{*})$ which satisfies:

$$1) \qquad \pi_{*}(\eta_{\phi}) = \xi_{\phi} \qquad 6.6$$

$$2) \qquad \eta_{\phi} \lrcorner\, \alpha = \phi \circ \pi \qquad 6.7$$

That η_{ϕ} is, in fact, an element of $\varepsilon(L,\nabla)$ follows by direct computation: first, it is clear that η_{ϕ} is $\mathbb{C}^{*}$-invariant since α itself is $\mathbb{C}^{*}$ invariant. Secondly, one has[15]:

$$1) \qquad \mathcal{L}_{\eta_{\phi}} \alpha = d(\eta_{\phi} \lrcorner\, \alpha) + \eta_{\phi} \lrcorner\, d\alpha$$

$$= d(\phi \circ \pi) + \eta_{\phi} \lrcorner\, \pi^{*}(\omega)$$

$$= \pi^{*}(d\phi + \xi_{\phi} \lrcorner\, \omega) = 0 \qquad 6.8$$

$$2) \qquad \eta_{\phi} H = 2\pi i\, H\, (\eta_{\phi} \lrcorner\, (\alpha - \bar{\alpha})) = 0 \qquad 6.9$$

so that η_{ϕ} also has the second defining property of $\varepsilon(L,\nabla)$.

Next, it must be shown that the map $\phi \longmapsto \eta_{\phi}$ preserves brackets. Again, a direct computation gives:

$$\left.\begin{aligned} &(1)\quad \pi_*([\eta_\phi,\eta_\chi]) = [\xi_\phi,\xi_\chi] = \xi_{[\phi,\chi]} && 6.10 \\ &(2)\quad [\eta_\phi,\eta_\chi] \lrcorner\, \alpha = \eta_\phi(\eta_\chi \lrcorner\, \alpha) \\ &\qquad = (\xi_\phi(\chi)) \circ \pi \\ &\qquad = [\phi,\chi] \circ \pi \end{aligned}\right\} \quad \phi,\chi \in C^\infty_{\mathbb{R}}(M) \qquad 6.11$$

Finally, since it is clear that the map $C^\infty_{\mathbb{R}}(M) \to \varepsilon(L,\nabla)$: $\phi \longmapsto \eta_\phi$ is one to one, all that remains to be shown is that it is also surjective. To do this, suppose that $\eta \in \varepsilon(L,\nabla)$ is given. Since η and α are $\mathbb{C}^*$ invariant, the function $\eta \lrcorner\, \alpha \in C^\infty(L^*)$ must be of the form:

$$\eta \lrcorner\, \alpha = \phi \circ \pi \qquad 6.12$$

for some $\phi \in C^\infty(M)$. But, $\eta\, H = 0$, so that:

$$\eta \lrcorner (\alpha - \bar{\alpha}) = 0, \qquad 6.13$$

that is, ϕ is real. Also $\mathcal{L}_\eta\, \alpha = 0$, so one has:

$$d(\eta \lrcorner\, \alpha) + \eta \lrcorner\, d\alpha = 0 \qquad 6.14$$

Hence:

$$d\phi + (\pi_*\, \eta) \lrcorner\, \omega = 0 \qquad 6.15$$

at each point of M. Thus $\pi_*(\eta) = \xi_\phi$ and $\eta = \eta_\phi$, and the map $\phi \longmapsto \eta_\phi$ is indeed an isomorphism of Lie algebras.

These results can be summarized in a commutative diagram of Lie algebra homomorphisms:

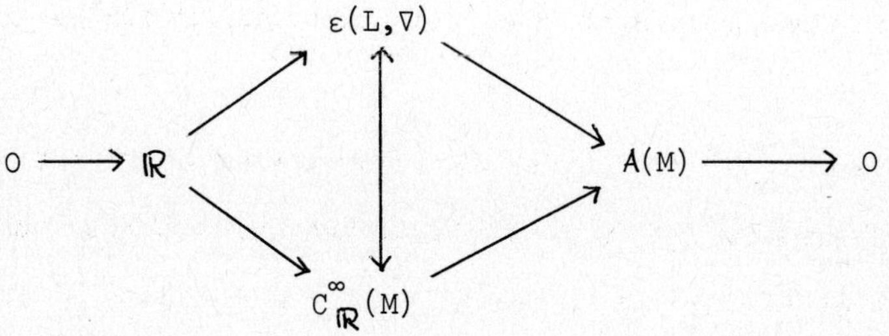

Here the vertical arrow is the natural isomorphism $\phi \longmapsto \eta_\phi$. The upper and lower sequences are exact (that is the kernel of each map is the image of the previous map) and the lower sequence is the real part of the one mentioned in §3. In the upper sequence, the map $\mathbb{R} \to \varepsilon(L,\nabla)$ sends each $r \in \mathbb{R}$ to the vertical vector field η_r which generates the flow:

$$L^* \times \mathbb{R} \to L^* : (\ell,t) \longmapsto e^{2\pi i\, rt}\, \ell \;.$$

The map $\varepsilon(L,\nabla) \to A(M)$ is simply the projection π_*.

It is now quite simple to see how a classical observable $\phi \in C^\infty_{\mathbb{R}}(M)$ acts as an operator on $\Gamma(L)$ and hence on $\mathcal{H}$. Each section $s \in \Gamma(L)$ defines a function $f_s : L^* \to \mathbb{C}$ according to:

$$f_s(\ell).\ell = s(m) \;;\quad m \in M,\quad \ell \in L^*_m \;. \qquad 6.16$$

Conversely, any function $f : L^* \to \mathbb{C}$ with the homogeneity property:

$$f(z.\ell) = z^{-1}.f(\ell) \; ; \quad \ell \in L^*, \quad z \in \mathbb{C}^* \qquad 6.17$$

defines a section $s_f \in \Gamma(L)$ (via eqn. 6.16). Thus[16] $\Gamma(L)$ can be identified with the subspace of $C^\infty(L^*)$ of functions satisfying eqn. 6.17.

Now if $\phi \in C^\infty_{\mathbb{R}}(M)$ then η_ϕ is $\mathbb{C}^*$-invariant and, for any section $s \in \Gamma(L)$, the function $\eta_\phi f_s \in C^\infty(L^*)$ satisfies eqn 6.17 and thus defines a section[17] $-2\pi i.\ \delta_\phi\ s \in \Gamma(L)$. The map $\delta_\phi : \Gamma(L) \to \Gamma(L)$ is linear and Hermitian (since η_ϕ is real); it is called the <u>quantum operator</u>[18] corresponding to ϕ.

It follows from the fact that $\phi \longmapsto \eta_\phi$ is a Lie algebra isomorphism that:

$$\left.\begin{array}{lll} 1) & \delta_{z\phi} = z.\delta_\phi & \\ 2) & \delta_{(\phi+\chi)} = \delta_\phi + \delta_\chi & \\ 3) & -2\pi i.\delta_{[\phi,\chi]} = [\delta_\phi, \delta_\chi] & \end{array}\right\} \phi, \chi \in C^\infty_{\mathbb{R}}(M),\ z \in \mathbb{C} \qquad \begin{array}{l} 6.18 \\ 6.19 \\ 6.20 \end{array}$$

Thus the map δ which sends a classical observable ϕ to the corresponding quantum operator δ_ϕ defines a representation of the Lie algebra $C^\infty_{\mathbb{R}}(M)$.

To summarize, each classical observable $\phi \in C^\infty_{\mathbb{R}}(M)$ is represented as a vector field η_ϕ on L^*. One recovers ϕ by contracting η_ϕ with α and by projecting η_ϕ into M one obtains the generator ξ_ϕ of the corresponding family of canonical transformations. Finally, ϕ emerges in its role as a quantum operator through the natural action of η_ϕ on the sections of L.

Example: Suppose that $M = T^*X$ is the cotangent bundle of a configuration space X. The canonical 2-form ω on M is exact, so that (M,ω) is quantizable: in fact, the line bundle (L,π,M) is simply the trivial bundle (see appendix A) and, when the space of sections $\Gamma(L)$ is identified with $C^\infty(M)$, the connection ∇ is given by:

$$\nabla_\xi \chi = \xi\chi + 2\pi i(\xi \lrcorner \theta) \cdot \chi \; ; \quad \xi \in \mathcal{U}(M), \quad \chi \in C^\infty(M) \qquad 6.21$$

where θ is the canonical 1-form. The corresponding connection form on $L^* = M \times \mathbb{C}^*$ is:

$$\alpha = \theta + \frac{1}{2\pi i} \cdot \frac{dz}{z} \qquad 6.22$$

and the vector field $\eta_\phi \in \mathcal{U}(M \times \mathbb{C}^*)$ generated by a classical observable $\phi \in C^\infty_{\mathbb{R}}(M)$ is:

$$\eta_\phi = \xi_\phi + 2\pi i\, z.(\phi - A) \frac{\partial}{\partial z} - 2\pi i\, \bar{z} \cdot (\phi - A) \frac{\partial}{\partial \bar{z}} \qquad 6.23$$

where A is the 'action' function of ϕ :

$$A = \xi_\phi \lrcorner \theta \in C^\infty_{\mathbb{R}}(M) \quad . \qquad 6.24$$

For future reference[19], note that the integral curves of η_ϕ are of the form:

$$t \longmapsto (\gamma(t),\ z_o \exp(2\pi i \int_o^t (\phi - A)\, dt)) \in M \times \mathbb{C}^*$$

where γ is an integral curve of ξ_ϕ and the integration is along γ. □.

In the form in which it is defined, δ_ϕ is not very useful in explicit calculations: it is more convenient to express it directly as an operator on $\Gamma(L)$. This is done as follows: suppose that $\phi \in C^\infty_{\mathbb{R}}(M)$ and that $s \in \Gamma(L)$ is a section which does not vanish in some open set $U \subset M$. Then $s(U)$ is a smooth surface in $L^* \cap \pi^{-1}(U)$ and the vector field:

$$\zeta_\phi = \eta_\phi - s_* \pi_* \eta_\phi = \eta_\phi - s_* \xi_\phi \qquad 6.25$$

on $s(U)$ is vertical, that is, it is tangent to the fibres in L^*. Also:

$$\zeta_\phi f_s = \eta_\phi f_s \text{ on } s(U) \qquad 6.26$$

since f_s is constant on $s(U)$.

Now choose $m \in U$ and consider the map:

$$\lambda : \mathbb{C}^* \to L_m^* : z \longmapsto z \,.\, s(m) \quad .$$

Since $(\zeta_\phi)_{s(m)}$ is tangent to L_m^*, one has:

$$(\zeta_\phi)_{s(m)} = \lambda_*\left(y \frac{\partial}{\partial z} + \bar{y} \frac{\partial}{\partial \bar{z}}\right)_{z=1} \qquad 6.27$$

for some $y \in \mathbb{C}$. But, for each $z \in \mathbb{C}^*$:

$$z = (f_s(\lambda(z)))^{-1} \qquad 6.28$$

so that

$$y = (\zeta_\phi(f_s^{-1}))\ (s(m))$$

$$= -\ (\zeta_\phi(f_s))\ (s(m))$$

$$= -\ (\eta_\phi(f_s))\ (s(m)) \qquad 6.29$$

Now the map λ is linear, so:

$$\lambda^* \alpha = \frac{1}{2\pi i} \cdot \frac{dz}{z}$$

by the second defining property of the connection 1-form. Thus y is also given by:

$$y = 2\pi i\ (\zeta \lrcorner\ \alpha)\ (s(m)) \qquad 6.30$$

Hence, on s(U):

$$\eta_\phi(f_s) = -2\pi i(\zeta_\phi \lrcorner \alpha)$$

$$= -2\pi i(\eta_\phi - s_* \xi_\phi) \lrcorner \alpha$$

$$= 2\pi i(\xi_\phi \lrcorner s^* \alpha - \phi) \circ \pi \qquad 6.31$$

implying:

$$-2\pi i.\delta_\phi s = \nabla_{\xi_\phi} s - 2\pi i\phi.s \qquad 6.32$$

in U. By continuity, this formula is also true sections which do vanish in U. Thus the identity:

$$-2\pi i.\delta_\phi = \nabla_{\xi_\phi} - 2\pi i.\phi \qquad 6.33$$

holds globally.

Prequantization is thus a general procedure for constructing a representation of the Lie algebra $C^\infty_{\mathbb{R}}(M)$ by Hermitian operators on $\Gamma(L)$ and hence on the Hilbert space $\mathcal{H}$. However the construction is, as yet, incomplete. This is best illustrated by the example introduced above which is central to the physical interpretation of the theory.

Consider a classical system consisting of a free particle moving in a configuration space X; the corresponding phase space is the cotangent bundle $M = T^*X$ of X. In this system, a distinguished role is played by the subalgebra $\mathcal{L}$ of $C^\infty_{\mathbb{R}}(M)$ consisting of functions which are linear in momentum, that is of functions $\chi \in C^{\infty}_{\mathbb{R}}(M)$ of the form:

$$\chi(m) = \zeta_x \lrcorner\, p + f(x); \quad (x,p) = m \in M, \quad x \in X, \qquad 6.34$$

where ζ is a real vector field on X and f is a smooth real function on X.

This subalgebra is spanned by the generators of:

1) The transformations of the phase space T^*X which leave the fibres invariant
2) The transformations of T^*X induced by infinitesimal diffeomorphisms of the configuration space X.

The Dirac problem, which was originally stated for systems in which X had a natural linear structure, can be formulated in this general context as the problem of finding a Hilbert space representation of the Lie algebra of classical observables (or of some suitable subalgebra) such that L is represented irreducibly (that is, no subspace of the Hilbert space is invariant under its action). A representation satisfying this condition will provide a kinematical model for the underlying quantum system.

The irreducibility condition reflects the fact that, in the classical system, L generates transformations of the phase space under which any state (point in M) can be mapped into any other nearby state. The corresponding quantum system should have an analogous property: in physical terms, no linear subspace of the quantum phase space (the set of rays in the Hilbert space) should be invariant with respect to changes of position and momentum.

Now on applying the quantization procedure one obtains a representation of $C^{\infty}_{\mathbb{R}}(M)$ by operators on $C^{\infty}(M)$ (which is isomorphic with $\Gamma(L)$ in this case). However, L is not represented irreducibly: in particular the subspace of $C^{\infty}(M)$ of functions which are constant on the fibres of $T^{*}X = M$ (that is, of functions which are independent of momentum) is invariant under the action of L, and is maximal with respect to this condition.

This example both illustrates the problem and suggests its solution. One must construct the representation space not from the whole of $C^{\infty}(M)$ but from the subspace of functions which are constant on the fibres: that is from the wave functions on configuration space rather than phase space. Though this will restrict the class of observables which can be quantized, it will result in an irreducible representation of L.

A similar problem arises in other situations, in particular when the classical system admits a transitive symmetry (or invariance) group, that is, a Lie group G which acts on the phase space as a group of canonical transformations without leaving any subset invariant. Such a system is called an elementary system for the group. (For example, a free particle in Minkowski space is an elementary system for the Poincaré group). When the prequantization procedure is carried out, the Lie group reappears as a group of unitary transformations of the quantum Hilbert space (this will be discussed further in §8): that is, one obtains a unitary representation of the group. Unfortunately, the resulting quantum system will not, in general, be an elementary system for the group, in the sense that the group representation will be reducible: it will be possible to express the Hilbert space as the direct sum of two or more G-invariant subspaces. The quantum system will be a composite system for the group: it will split up into a number of subsystems each admitting G as a symmetry group.

Fortunately, it is often possible to obtain irreducible representations by choosing a suitable polarization of phase space. The defining properties of a polarization allow it to play a role analogous to that played by the fibres of T^*X in the first example.

7. Quantization

Definition: A polarization of a symplectic manifold (M,ω) is a map P which assigns to each point $m \in M$ a subspace $P_m \subset (T_m M)^{\mathbb{C}}$ of the complexified tangent space at m, satisfying:

(1) P is involutory: that is, the set of vector fields $\mathcal{U}_P(M) = \{\xi \in \mathcal{U}(M) \mid \xi_m \in P_m \ \forall\ m \in M\}$ is closed under the Lie bracket.

(2) P is smooth: $\forall\ m \in M$, $P_m = \{\xi_m \mid \xi \in \mathcal{U}_P(M)\}$.

(3) P is maximally isotropic: $\forall\ m \in M$, $\omega(P_m, P_m) = 0$ and no other subspace of $(T_m M)^{\mathbb{C}}$ which contains P_m has this property. (In particular, if M is $2n$ dimensional, then P_m is n dimensional).

(4) $\forall\ m \in M$, $D_m^{\mathbb{C}} = P_m \cap \bar{P}_m$ has constant dimension k.

Two examples have already been encountered. In the first $k = n$, in the second $k = 0$:

I) A cotangent bundle $M = T^*X$ has a natural polarization given by:

$$P_m = D_m^{\mathbb{C}} ; \quad m \in M \tag{7.1}$$

where $D_m^{\mathbb{C}}$ is the complexified tangent space to the fibre through m; P is called the vertical polarization.

The defining properties of a polarization are established for P

as follows: a vector field $\xi \in \mathcal{U}_P(M)$ is characterized by the condition:

$$\xi(f \circ pr) = 0 \quad \forall \quad f \in C^\infty(X)$$

and has the property:

$$\xi \lrcorner \theta = 0 \qquad 7.3$$

Here pr: $M = T^* X \to X$ is the natural projection and $\theta \in \Omega^1(M)$ is the canonical 1-form. Thus if ξ, $\zeta \in \mathcal{U}_P(M)$ then:

$$[\xi,\zeta]\ (f \circ pr) = \xi(\zeta(f \circ pr)) - \zeta(\xi(f \circ pr)) = 0 \quad \forall\ f \in C^\infty(X) \qquad 7.4$$

so that $[\xi,\zeta] \in \mathcal{U}_p(M)$, that is, P is involutory. Also, P is isotropic since if $\xi,\zeta \in \mathcal{U}_P(M)$ then:

$$\omega(\xi,\zeta) = \xi(\zeta \lrcorner \theta) - \zeta(\xi \lrcorner \theta) - [\xi,\zeta] \lrcorner \theta \ = 0 \qquad 7.5$$

(using $\omega = d\theta$). Finally, P is maximally isotropic since $D_M^{\mathbb{C}}$ is n-dimensional over $\mathbb{C}$ and smooth since the fibres (that is, the cotangent spaces) are smooth submanifolds of $M = T^* X$.

One can introduce local canonical coordinates $\{x^a, p_a\}$ into M by lifting local coordinates $\{x^a\}$ from X (see §1); $D_m^{\mathbb{C}}$ is then spanned by the set $\{\partial/\partial p_a\}$ and that P is isotropic is simply another way of saying that each of the Lagrange brackets:

$$\tfrac{1}{2}\{\partial/\partial p_a, \partial/\partial p_b\} = \omega(\partial/\partial p_a, \partial/\partial p_b) \qquad 7.6$$

vanishes identically.

This polarization is real in the sense that for each $m \in M$, $P_m = \bar{P}_m$ and $k = n$.

II) A Kähler manifold (M,ω,J) has a natural polarization defined by:

$$P_m = \{\xi_m \in (T_m M)^{\mathbb{C}} \mid J_m \xi_m = i\,\xi_m\} \quad ; \qquad m \in M . \qquad 7.7$$

In local complex coordinates $\{z^a\}$, P_m is the linear span of the set $\{\partial/\partial \bar{z}^a\}$ of antiholomorphic coordinate vectors at m.

This example is extreme in the sense that $P_m \cap \bar{P}_m = \{0\}$ for each $m \in M$. A polarization with this property is called a <u>Kähler polarization</u>.

In the case where the phase space (M,ω) is the cotangent bundle T^*X of a configuration space X, the idea is to construct the Hilbert space of the underlying quantum system out of the functions on M which are constant on the integral surfaces of the vertical polarization P (that is, on the fibres in T^*X). These can be thought of as wave functions on the base space X, which is naturally identified with the factor space M/P (the space of integral surfaces).

Unfortunately, a function on T^*X which is constant on the fibres will not be square integrable with respect to the natural volume element ω^n (unless, of course, it is zero) so that the pre-Hilbert space structure used in prequantization is of no use here.

The direction in which one should proceed is suggested by a re-examination of the Schrödinger prescription, which is applicable in the special case where X is Euclidean space (and so has a preferred class of coordinate systems). Here, the wave functions are most naturally regarded not as functions on X but as square integrable $^1/_2$-densities. To be explicit, let $B^*X^{\mathbb{C}}$ denote the frame bundle of $T^*X^{\mathbb{C}}$: that is, each point of $B^*X^{\mathbb{C}}$ is an (n + 1)-tuple $(x, \alpha_1, \ldots, \alpha_n)$ consisting of a point $x \in X$ and a basis $\{\alpha_1, \alpha_2, \ldots, \alpha_n\}$ for the complexified cotangent space $T^*_xX^{\mathbb{C}}$ at x (as a vector space over $\mathbb{C}$). In the language of fibre bundle theory, $B^*X^{\mathbb{C}}$ is the associated principal bundle of $T^*X^{\mathbb{C}}$: its structure group is the general linear group $GL(n,\mathbb{C})$.

An r <u>density</u> (r is a real number) on X is a smooth function

$$\phi : B^*X \to \mathbb{C}$$

which transforms under the action of $GL(n,\mathbb{C})$ on $B^*X^{\mathbb{C}}$ (by right translation) according to:

$$\phi \circ g = |\Delta_g|^{-r}.\phi \; ; \quad g : B^*X^{\mathbb{C}} \to B^*X^{\mathbb{C}}; \quad g \in GL(n,\mathbb{C}) \qquad 7.8$$

where Δ_g is the determinant of g. If r = 1 then ϕ is simply a density.

There is a natural inner product <. , .> on the space of $^1/_2$ - densities. If ϕ and ψ are $^1/_2$ - densities, then the product $\phi.\bar{\psi}$ is a density; $<\phi,\psi>$ is defined by integrating[20] $\phi.\bar{\psi}$ over X. The quantum phase space corresponding to the configuration space X is modelled on the pre-Hilbert space formed by the square integrable $^1/_2$-densities (that is, the $^1/_2$-densities ϕ for which $<\phi,\phi>$ is finite). Of course, when X is Euclidean space, there is a natural volume element which can be used to identify $^1/_2$-densities with functions in $C^\infty(X)$. But in the case of a general configuration space X, there will be no such natural identification; it is then more useful to think of the quantum wave functions as $^1/_2$-densities (which have a natural inner product) rather than as smooth functions (which do not).

To make use of this idea when M is not a cotangent bundle, or when P is not the vertical polarization, one must first lift the construction to T^*X, that is, one must identify the $^1/_2$-densities on X with a suitable class of objects on T^*X. This is done by introducing the concept of a $^1/_2$-P-density:

Let B_m^P denote the set of all bases for $P_m \subset (T_m(T^*X))^{\mathbb{C}}$ at $m = (x,p) \in T^*X$ (X is now a general configuration space) and put:

$$B^P(T^*X) = \bigcup_{m \in T^*X} \{m\} \times B_m^P \quad ;$$

$B^P(T^*X)$ is called the frame bundle of P. It is a principal $GL(n,\mathbb{C})$ bundle. A $^1/_2$-P-density is a function

$$\nu : B^P(T^*X) \to \mathbb{C}$$

which transforms under the action of $GL(n,\mathbb{C})$ according to:

$$\nu \circ g = |\Delta_g|^{-\frac{1}{2}}.\nu \; ; \quad g : B^P(T^*X) \to B^P(T^*X); \quad g \in GL(n,\mathbb{C}) \qquad 7.9$$

Each basis $\alpha_1,\alpha_2,\ldots,\alpha_n \in B^*_x X^{\mathbb{C}}$ at $x \in X$ defines a basis $\xi_1,\ldots,\xi_n$ at $m = (x,p)$ in the fibre above x, specified by:

$$\xi_i \lrcorner \, \omega + pr^*(\alpha_i) = 0 \quad . \qquad 7.10$$

Thus, each $^1/_2$-density μ on X defines a $^1/_2$-P-density ν_μ on T^*X, given by:

$$\nu_\mu(m, \xi_1, \ldots, \xi_n) = \mu(x, \alpha_1,\ldots, \alpha_n) \qquad 7.11$$

However, not every $^1/_2$-P-density can be identified with a $^1/_2$-density on X; those that can be have the property that they are constant on the fibres in T^*X in a sense which needs some explanation.

Suppose that the n complex valued functions $z_1, z_2, \ldots, z_n \in C^\infty(X)$ have gradients which are linearly independent (over $\mathbb{C}$) in some open set $U \subset X$. Then the functions:

$$\phi_i = z_i \circ pr \in C^\infty(T^*X) \qquad 7.12$$

are constant on the fibres of T^*X, so that the corresponding Hamiltonian vector fields $\xi_{\phi_1} \ldots \xi_{\phi_n}$ are tangent to the fibres. Hence if $m = (x,p) \in pr^{-1}(U)$ then $(\xi_{\phi_1},\ldots \xi_{\phi_n})_m$ is a basis in B^P_m and, if μ is a $^1/_2$-density on X, then the corresponding $^1/_2$-P-density

ν_μ is given by:

$$\nu_\mu(m,(\xi_{\phi_1},\dots \xi_{\phi_n})_m) = \mu(x,(dz_1,\dots,dz_n)_x) \qquad 7.13$$

Thus the function

$$m \longmapsto \nu_\mu(m,(\xi_{\phi_1},\dots \xi_{\phi_n})_m)$$

is constant on the fibres of T^*X.

Conversely, a given $^1/_2$-P-density ν will be of the form $\nu = \nu_\mu$ for some $^1/_2$-density on X if it has the property: for any open set $V \subset T^*X$ and for any collection $\{\xi_1,\dots \xi_n\}$ of locally Hamiltonian vector fields in $\mathcal{U}_P(T^*X)$ which are linearly independent in V, the function:

$$V \to \mathbb{C} \; : m \longmapsto \nu(m,(\xi_1)_m,\dots (\xi_n)_m)$$

is constant on the fibres of T^*X. This follows by reversing the argument above since, locally, any set of vector fields $\{\xi_1,\dots,\xi_n\}$ with these properties must be of the form:

$$\xi_i = \xi_{\phi_i} \; ; \quad \phi_i \in C^\infty(T^*X) \qquad 7.14$$

where:

$$\phi_i = z_i \circ pr \; ; \quad z_i \in C^\infty(X). \qquad 7.15$$

This condition can be put into a more concise form by introducing the Lie derivative of a $^1/_2$-P-density. The definition of this needs a little care and, as the concept is needed later, I will explain it in the general context of a 2n-dimensional symplectic manifold (M,ω) with a polarization P. Here the frame $B^P(M)$ of P can be introduced in precisely the same way as before. It is helpful to think of $B^P(M)$ as a submanifold of the bundle $E(M)^{\mathbb{C}}$ of n-frames: each point of $E(M)^{\mathbb{C}}$ is an (n + 1)-tuple $(m, \zeta_1,\ldots, \zeta_n)$ consisting of a point $m \in M$ and n linearly independent vectors $\zeta_1 \ldots \zeta_n \in (T_m M)^{\mathbb{C}}$. Now any (real or complex) vector field $\eta \in \mathcal{U}(M)$ lifts naturally to a vector field $\tilde{\eta}$ on $E(M)^{\mathbb{C}}$. If η is real then $\tilde{\eta}$ is the tangent vector field to the flow on $E(M)^{\mathbb{C}}$ defined by dragging n-frames along η. When η is complex, $\tilde{\eta}$ is defined by lifting its real and imaginary parts separately. Explicitly, $\tilde{\eta}$ is given as follows. Let $\{\xi_1 \ldots \xi_n\} \subset \mathcal{U}(M)$ be vector fields, linearly independent in some open set $U \subset M$: these define a local section:

$$\sigma : U \to E(M)^{\mathbb{C}} : m \longmapsto (m, (\xi_1)_m, \ldots (\xi_n)_m)$$

of $E(M)$ (the condition that $\sigma(U) \subset B^P(M)$ is that $\xi_1,\ldots, \xi_n$ should belong to $\mathcal{U}_P(M)$). The value of $\tilde{\eta}$ on $\sigma(U) \subset E(M)^{\mathbb{C}}$ is given by:

$$\tilde{\eta} = \sigma_*(\eta) + v \qquad 7.16$$

where v is the vertical[21] tangent vector to the flow

$E(M)^{\mathbb{C}} \times \mathbb{R} \to E(M)$:

$$((m, \zeta_1 \ldots \zeta_n), t) \longmapsto (m, \zeta_1 + t[\xi_1, \eta], \ldots, \zeta_n + t[\xi_n, \eta])$$

If, therefore, η preserves P in the sense[22]:

$$\mathcal{L}_\eta \xi \in \mathcal{U}_P(M) \qquad \forall \ \xi \in \mathcal{U}_P(M)$$

(for example, η might be in $\mathcal{U}_P(M)$) then $\tilde{\eta}$ is tangent to $B^P(M) \subset E(M)^{\mathbb{C}}$ and η lifts to a vector field on $B^P(M)$. The **Lie** <u>derivative</u> along η of a $^1/_2$-P-form $\nu : B^P(M) \to \mathbb{C}$ is the $^1/_2$-P-form:

$$\mathcal{L}_\eta \nu = \tilde{\eta} \nu \qquad 7.17$$

Returning to the case $M = T^*X$, if $\xi_1 \ldots \xi_n \in \mathcal{U}_P(T^*X)$ are locally Hamiltonian, then:

$$\mathcal{L}_\eta \xi_i = 0 \qquad 7.18$$

for every locally Hamiltonian vector field $\eta \in \mathcal{U}_P(T^*X)$. Thus the $^1/_2$-P-densities ν which are of the form $\nu = \nu_\mu$ for some $^1/_2$-density μ on X are precisely those which satisfy

$$\mathcal{L}_\eta \nu = 0 \qquad 7.19$$

for every locally Hamiltonian vector field $\eta \in \mathcal{U}_P(T^*X)$.

In terms of the geometry of phase space, therefore, the wave functions of the Schrödinger prescription are the $^1/_2$-P-densities which are constant on the integral surfaces of the vertical polarization in the sense of eqn. 7.19.

By analogy, in the general case, the Hilbert space of a quantizable symplectic manifold should be constructed, not as in prequantization, from the sections of the line bundle L alone but from the products of sections of L with the $^1/_2$-P-densities of some polarization P: a subset of these objects which are constant in the directions in P will have a natural pre-Hilbert space structure.

This scheme can be (and has been) carried through. However, rather more fruitful results are obtained using a slight variation in which objects called $^1/_2$-P-forms are used in place of $^1/_2$-P-densities[23]. In the simplest cases (such as when the phase space is the cotangent bundle of an oriented configuration space) this is a technicality which can be largely ignored. However it does play an important role in the more complex systems.

Essentially, a $^1/_2$-P-form on a symplectic manifold (M,ω) with a polarization P, is a function $\nu : B^P(M) \to \mathbb{C}$ which transforms under right translation by $GL(n,\mathbb{C})$ according to:

$$\nu \circ g = (\Delta_g)^{-\frac{1}{2}} \nu \; ; \quad g : B^P(M) \to B^P(M); \quad g \in GL(n,\mathbb{C}) \quad .$$

To make this precise, one must remove the ambiguity in the

square root. This is done by replacing the general linear group by its double cover, the metalinear group: one takes the "square root of the group" rather than of the determinant.

Explicitly, the metalinear group $ML(n,\mathbb{C})$ is the subgroup of $GL(n+1,\mathbb{C})$ of matrices of the form:

$$\begin{bmatrix} g & 0 \\ 0 & z \end{bmatrix}$$

where $g \in GL(n,\mathbb{C})$ and $z^2 = \Delta_g$. The covering map $\sigma: ML(n,\mathbb{C}) \rightarrow GL(n,\mathbb{C})$ is defined by:

$$\sigma\left(\begin{bmatrix} g & 0 \\ 0 & z \end{bmatrix}\right) = g \qquad 7.20$$

It is a double cover since

$$\sigma^{-1}(g) = \begin{bmatrix} g & 0 \\ 0 & \pm z \end{bmatrix} \quad ; \quad g \in GL(n,\mathbb{C}); \quad z = (\Delta_g)^{\frac{1}{2}} \quad . \qquad 7.21$$

There is also a natural group homomorphism $\chi : ML(n,\mathbb{C}) \rightarrow \mathbb{C}^*$ (the multiplicative group of complex numbers) given by:

$$\chi\left(\begin{bmatrix} g & 0 \\ 0 & z \end{bmatrix}\right) = z$$

This gives rise to the commutative diagram:

$$\begin{array}{ccc} ML(n,\mathbb{C}) & & \\ \downarrow & \searrow^{\chi^2} & \\ & & \mathbb{C}^* \\ \downarrow & \nearrow_{\Delta_g} & \\ GL(n,\mathbb{C}) & & \end{array}$$

so that χ is a well defined 'square root of the determinant'.

the $^1/_2$-P-forms are functions defined not on the frame bundle $B^P(M)$, but on a double covering $\tilde{B}^P(M)$, called a metalinear frame bundle. To be precise a metalinear frame bundle pr: $\tilde{B}^P(M) \to M$ for P is a principal $ML(n,\mathbb{C})$ bundle together with a covering map $\rho : \tilde{B}^P(M) \to B^P(M)$ which makes the diagram

$$\begin{array}{ccc} \tilde{B}^P(M) \times ML(n,\mathbb{C}) & \longrightarrow & \tilde{B}^P(M) \\ \downarrow & & \downarrow \\ B^P(M) \times GL(n,\mathbb{C}) & \longrightarrow & B^P(M) \end{array}$$

commute. The horizontal arrows are the natural group actions (right translation), the second vertical arrow is the covering map ρ and the first is the product of ρ with σ.

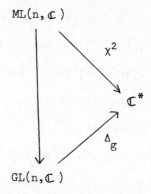

There is no guarantee that $\tilde{B}^P(M)$ exists and, even when it does exist, it will not, in general, be unique. The existence condition is that a certain class in $H^2(M,\mathbb{Z}_2)$ associated with P (the 'obstruction') should vanish; when it does vanish, the various possible choices for $\tilde{B}^P(M)$ are parameterized by the cohomology group $H^1(M, \mathbb{Z}_2)$. This is explained in detail in appendix B.

Though the construction of $\tilde{B}^P(M)$ given in the appendix is rather abstract, one can gain some geometrical insight into its meaning by looking at the local trivializations of $B^P(M)$. A local trivialization of $B^P(M)$ is simply a set of vector fields $\{\xi_1, \ldots \xi_n\} \subset \mathcal{U}(M)$ which form a basis for P_m at each point m of some open set $U \subset M$. Each element $(m, \zeta_1, \ldots \zeta_n)$ of $B^P(M)|U$ is then represented by a pair:

$$(m, g)$$

where g is the $n \times n$ complex matrix in $GL(n, \mathbb{C})$ defined by:

$$\zeta_j = (\xi_i)_m \, g_{ij} \tag{7.23}$$

Associated with $(m, \zeta_1 \ldots \zeta_n)$, there are two points of $\tilde{B}^P(M)$ represented by the pairs:

$$\left(m, \begin{bmatrix} g & 0 \\ 0 & z \end{bmatrix}\right) \quad \text{and} \quad \left(m, \begin{bmatrix} g & 0 \\ 0 & -z \end{bmatrix}\right)$$

where z is one of the square roots of Δ_g. One can think of these two pairs as the two 'meta-frames' corresponding to $(m, \zeta_1 \ldots \zeta_n)$ (there is a close analogy in general relativity where there are two spin frames corresponding to a given null tetrad).

The subtlety arises when one replaces $(\xi_1, \ldots \xi_n)$ by a second set of vector fields $(\tilde{\xi}_1, \ldots \tilde{\xi}_n)$ given by:

$$(\xi_j)_m = (\tilde{\xi}_i)_m \cdot h_{ij}(m) \qquad 7.24$$

where $h: M \to GL(n, \mathbb{C})$ is some smooth map. Then the pair (m,g) representing $(m, \zeta_1, \ldots, \zeta_n)$ transforms according to:

$$(m,g) \longmapsto (m, h(m)g) \quad .$$

However, what is less clear is which of the two pairs:

$$(m, \begin{bmatrix} hg & 0 \\ 0 & wz \end{bmatrix}) \quad \text{and} \quad (m, \begin{bmatrix} hg & 0 \\ 0 & -wz \end{bmatrix}) \;; \quad w^2 = \Delta_h$$

represents in the new local trivialization the same metalinear frame as

$$(m, \begin{bmatrix} g & 0 \\ 0 & z \end{bmatrix})$$

did in the old.

The vanishing of the obstruction is simply a precise formulation of the condition that a consistent (but not necessarily unique) choice can be made here for every change of local trivialization, so that there is a well defined concept of a 'meta-frame'.

Having chosen a metalinear frame bundle $\tilde{B}^P(M)$ (that is, assuming that the obstruction does vanish) one defines a $^1/_2$-P-form to be any smooth function $\nu: \tilde{B}^P \to \mathbb{C}$ which transforms under the action of $ML(n,\mathbb{C})$ according to:

$$\nu \circ \tilde{g} = \chi(\tilde{g})^{-1} \cdot \nu \qquad 7.25$$

where $\tilde{g} : \tilde{B}^P(M) \to \tilde{B}^P(M)$ is right translation by $\tilde{g} \in ML(n,\mathbb{C})$.

Alternatively, one can regard the $^1/_2$-P-forms as sections of the line bundle $\pi: L^P \to M$ with bundle space:

$$L^P = \bigcup_{m \in M} \{m\} \times L^P_m$$

Here L^P_m is the set of functions $\nu: \tilde{B}^P_m \to \mathbb{C}$. ($\tilde{B}^P_m$ is the set of meta-frames at $m \in M$) which transform under the action of $ML(n,\mathbb{C})$ according to:

$$\nu(b\tilde{g}) = \chi(\tilde{g})^{-1} \cdot \nu(b); \quad b \in \tilde{B}^P_M, \quad \tilde{g} \in ML(n,\mathbb{C})$$

The elements of L^P_m are called $^1/_2$-P-forms at m.

The Hilbert space for the quantum system is to be constructed from the sections of[24] $L \otimes L^P$, where L is the line bundle used in prequantization, that is from the vector space $\Gamma(L \otimes L^P)$ of formal products

$$s.\nu \ ; \quad s \in \Gamma(L), \quad \nu \in \Gamma(L^P)$$

modulo the equivalence relation:

$$(\phi s).\nu = s.(\phi\nu) \; ; \quad \phi \in C^\infty(M)$$

Pursuing the analogy with the Schrödinger prescription, the first stage in this construction is to define the Lie derivative of a $^1/_2$-P-form along a vector field η which preserves P. Now η lifts naturally to a vector field as $B^P(M)$ (as before) and hence to a vector field η' on $\tilde{B}^P(M)$ (since the double covering $\tilde{B}^P(M) \to B^P(M)$ is a local diffeomorphism). Thus, as with $^1/_2$-P-densities, one can define[25]:

$$\mathcal{L}_\eta \nu = \eta' \nu \tag{7.26}$$

where $\nu : B^P(M) \to \mathbb{C}$ is a $^1/_2$-P-form.

The analogues of the wave functions of the Schrödinger prescription are thus the sections $\psi \in \Gamma(L \otimes L^P)$ which can be represented in the form $\psi = s.\nu$ where:

$$\mathcal{L}_\xi \nu = 0 \quad \text{and} \quad \nabla_\xi s = 0 \tag{7.27}$$

for every locally Hamiltonian vector field $\xi \in \mathcal{U}_P(M)$. The set of all these sections forms a complex vector space, denoted W^P.

The next step is to define an inner product on W^P: again pursuing the analogy with the Schrödinger case, the natural thing to do would be to factor out the integral surfaces of P and to define the inner product as an integral over M/P. This is all very well when the polarization is real (though it is still a strong additional requirement that M/P should be a manifold) but when P is complex, it does not have any integral manifolds even locally (for example, even though the anti-holomorphic vector fields on a Kähler manifold are in involution, they are not surface forming).

To overcome this[26], it is necessary to look again at the motivation for introducing $^1/_2$-P-forms in the first place: they were introduced because, in general, sections of L which are constant in the directions in P are not square integrable with respect to the natural volume element of M. However, in very crude terms, it is only the real directions in P which give the trouble and only these need be factored out. In the other directions, it is quite in order to integrate using the natural volume element.

In a more precise language, this is achieved as follows. For any polarization P, the map D which assigns to each point $m \in M$ the real subspace:

$$D_m = P_m \cap \bar{P}_m \cap T_m(M) \qquad 7.28$$

of T_mM is involutory. Also, by the fourth defining property of a polarization, D_m has constant real dimension k. Thus D is surface forming. It is now necessary to make the additional (very strong) assumption that these surfaces are simply connected and that the factor space M/D (the space of integral surfaces) is a

Hausdorff manifold: $X = M/D$ can be thought of as M with the real directions in P factored out. It is the analogue of configuration space in the Schrödinger case.

Each pair $\psi_1 = s_1 . \nu_1$, $\psi_2 = s_2 . \nu_2$ of 'wave functions' in W^P defines a density on M/D. Explicitly: let $m \in M$ and let $\tilde{b} \in \tilde{B}^P_m$ be a metaframe at m such that $\rho(\tilde{b}) = (\xi_1, \ldots, \xi_n) \in B^P_m$ and $\xi_1 \ldots \xi_k$ is a basis for D_m. Choose $\zeta_1, \ldots, \zeta_n \in T_m M^{\mathbb{C}}$ so that $\{\xi_1, \ldots \xi_n, \zeta_1 \ldots \zeta_n\}$ is a symplectic basis, that is so that:

$$\omega(\xi_i, \xi_j) = 0 = \omega(\zeta_i, \zeta_j) \; ; \qquad \omega(\xi_i, \zeta_j) = \delta_{ij} \qquad 7.29$$

Then $c = \{pr_*(\xi_{k+1}), \ldots, pr_*(\xi_n), pr_*(\zeta_1) \ldots pr_*(\zeta_n)\}$ will be a basis for $T_x X^{\mathbb{C}}$ at $x = pr(m)$ (here $pr: M \to X = M/D$ is the natural projection). The density (ψ_1, ψ_2) is defined by:

$$(\psi_1, \psi_2)\,(x, c^*)$$

$$= (s_1, s_2)(m) . | \omega^{n-k} (\xi_{k+1}, \bar{\xi}_{k+1}, \xi_{k+1}, \ldots, \xi_n, \bar{\xi}_n)|^{\frac{1}{2}} \nu_1(\tilde{b}) . \nu_2(\tilde{b}) \qquad 7.30$$

where $\omega^{n-k} = \omega \wedge \omega \wedge \ldots \wedge \omega$ and c^* is the dual basis to c.

It must be shown that (ψ_1, ψ_2) is a well defined density. First, suppose that $\tilde{b}$ is replaced by $\widetilde{bg}$, where $\tilde{g} \in ML(n; \mathbb{C})$. If $\widetilde{bg}$ is to have the same properties as $\tilde{b}$ then $\tilde{g}$ must be of the form:

$$\tilde{g} = \begin{bmatrix} a & \cdot & 0 \\ 0 & b & 0 \\ 0 & 0 & z \end{bmatrix} \qquad 7.31$$

where $a \in GL(k, \mathbb{R})$, $b \in GL(n-k, \mathbb{C})$ and $z^2 = \Delta_a \cdot \Delta_b$. Then c is replaced ch, where

$$h = \left[\begin{array}{c|c} b & 0 \\ \hline 0 & {}^T\begin{bmatrix} a & \cdot \\ 0 & b \end{bmatrix}^{-1} \end{array}\right] \in GL(2n-k, \mathbb{C}) \qquad 7.32$$

and h has determinant:

$$\Delta_h = \Delta_b \, \Delta_a^{-1} \, \Delta_b^{-1} = \Delta_a^{-1} \qquad 7.32a$$

But, from eqn. 7.30

$$(\psi_1, \psi_2)\,(x, (ch)^*)$$

$$= (\psi_1, \psi_2)\,(x, c^{*T} h^{-1})$$

$$= |\Delta_b \cdot \bar{\Delta}_b|^{\frac{1}{2}} \; \chi(\tilde{g})^{-1} \cdot \overline{\chi(\tilde{g})^{-1}} \cdot (\psi_1, \psi_2)\,(x, c^*)$$

$$= |\tfrac{1}{z}|^2 \cdot (\Delta_b \cdot \bar{\Delta}_b)^{\frac{1}{2}} \cdot (\psi_1, \psi_2)\,(x, c^*)$$

$$= |\Delta_a^{-1}| \; (\psi_1, \psi_2)\,(x, c^*) \qquad 7.33$$

so that (ψ_1,ψ_2) does indeed have the transformation properties of a density.

Secondly[26], it follows by extending $\xi_1,\ldots\ \xi_k$ to locally Hamiltonian vector fields in $\mathcal{U}_P(M)$ that the definition of (ψ_1,ψ_2) at $x \in X$ is independent of the point $m \in pr^{-1}(x)$ at which eqn. 7.30 is evaluated.

Finally, the inner product $<\psi_1,\psi_2>$ is defined by integrating (ψ_1,ψ_2) over X:

$$<\psi_1,\psi_2> = \int_X (\psi_1,\psi_2) \qquad 7.34$$

The subspace of W^P of wave functions ψ for which $<\psi\ ,\psi\ >$ is finite forms a pre-Hilbert space, denoted $\mathcal{H}_o^P$. The completion $\mathcal{H}^P$ of $\mathcal{H}_o^P$ will be taken as the quantum Hilbert space.

Unfortunately, in reducing the Hilbert space of prequantization in this way, one also reduces the class of observables which can be quantized. In fact, the only classical observables which can be dealt with directly are those which preserve P, that is, the observables $\phi \in C^\infty_{\mathbb{R}}(M)$ which satisfy[27]:

$$\pounds_{\xi_\phi} \eta \in \mathcal{U}_P(M) \quad \forall\ \eta \in \mathcal{U}_P(M)$$

(for example, in the Schrödinger case, ϕ must be at most linear in the momentum). The quantum operator $\tilde{\delta}_\phi$ corresponding to an observable ϕ satisfying this condition acts on W^P by:

$$\tilde{\delta}_\phi(s.\nu) = (\delta_\phi\, s).\nu - \frac{1}{2\pi i}.s.\, \mathcal{L}_{\xi_\phi}\nu \qquad 7.35$$

This is clearly Hermitian: the one parameter unitary group generated by $\tilde{\delta}_\phi$ corresponds to dragging s along η_ϕ in L^* and ν along ξ_ϕ' in $\tilde{\wedge}_B^P(M)$.

Before turning to the question of how observables which do not preserve the polarization are to be quantized, I will first deal with a closely related problem: to what extent is the quantization procedure independent of the choice of polarization? More formally: given two polarizations P_1 and P_2, is it possible to construct a unitary isomorphism $U: H^{P_1} \to H^{P_2}$ with the property that, for each real observable ϕ which preserves both P_1 and P_2, the corresponding quantum operators $\tilde{\delta}^1_\phi$ and $\tilde{\delta}^2_\phi$ are related by:

$$\tilde{\delta}^1_\phi = U^{-1}\, \tilde{\delta}^2_\phi\, U \qquad 7.36$$

For real polarizations, and in a purely formal sense, the answer is yes: ignoring problems of convergence and so forth, it is possible, under certain conditions, to write down an expression for U which generalizes the familiar Fourier transform between the p and q representations of elementary quantum mechanics. However, it must be emphasized that no complete answer is available and that this part of geometric quantization theory has yet to be put in its final form.

In detail, the idea is this: if P_1 and P_2 are real and transverse (that is, if they span the whole of T_mM at each point $m \in M$) then it is possible to construct a Hermitian 'pairing':

$$W^{P_1} \times W^{P_2} \to C^\infty(M) : (\psi_1, \psi_2) \longmapsto \psi_1 * \psi_2$$

which is linear in ψ_1, antilinear in ψ_2 and is related to the corresponding pairing of W^{P_2} with W^{P_1} by:

$$\psi_1 * \psi_2 = \overline{\psi_2 * \psi_1} \quad . \qquad 7.37$$

If W^{P_1} and W^{P_2} were finite dimensional, this would be sufficient to define a unique linear transformation:

$$U_{P_1P_2} : W^{P_1} \to W^{P_2}$$

satisfying:

$$\langle \psi_1, \psi_1 \rangle = \int_M \psi_1 * (U\psi_1) \cdot \omega^n \ \forall \psi_1 \in W^{P_1} \qquad 7.38$$

where $\langle\ ,\ \rangle$ is the inner product in W^{P_1}. In the infinite dimensional case, the question of whether or not $U_{P_1P_2}$ (it is called the BKS transform) exists is a difficult analytical problem which, in the present state of knowledge, can only be dealt with case by case[28].

The first step in the construction is to define a joint metalinear structure for P_1 and P_2. Since P_1 and P_2 are assumed to be transverse[29], each real basis $\{\xi_1, \ldots, \xi_n\} \in B_M^{P_1}$ at $m \in M$ extends uniquely to a real symplectic basis $\{\xi_1, \ldots, \xi_n, \ \zeta_1, \ldots, \zeta_n\}$ for T_mM in such a way that $\{\zeta_1, \ldots, \zeta_n\} \in B_M^{P_2}$. When $\{\xi_1, \ldots, \xi_n\}$ is replaced by $\{\tilde{\xi}_1, \ldots, \tilde{\xi}_n\}$, where:

$$\tilde{\xi}_j = \xi_i \, g_{ij} \; ; \quad g \in GL(n,\mathbb{C}) \qquad 7.39$$

$\{\zeta_1,\ldots,\zeta_n\}$ transforms into $\{\tilde{\zeta}_1,\ldots,\tilde{\zeta}_n\}$ where:

$$\tilde{\zeta}_j = \zeta_i \, {}^{T}g^{-1}_{ij} \qquad 7.40$$

Thus each basis in $B_M^{P_1}$ defines a basis in $B_M^{P_2}$ and conversely: $B^{P_1}(M)$ and $B^{P_2}(M)$ are naturally isomorphic and the choice of a metalinear frame bundle $\tilde{B}^{P_1}(M)$ for $B^{P_1}(M)$ automatically defines a metalinear frame bundle for $B^{P_2}(M)$.

Suppose, therefore, that a choice has been made for $\tilde{B}^{P_1}(M)$. Then each $1/2$-P_2-form ν_2 can be regarded as a function $\nu_2 : \tilde{B}^{P_1}(M) \to$ which transform under the action of $ML(n,\mathbb{C})$ (by right translation) according to:

$$\nu_2 \circ \tilde{g} = \chi(\tilde{g}) \; \nu_2 ; \quad \tilde{g} : B^{P_1}(M) \to B^{P_1}(M) ; \quad \tilde{g} \in ML(n,\mathbb{C}) \qquad 7.41$$

Thus, if ν_1 is a $1/2$-P_1-form, the quantity:

$$\nu_1(\tilde{b}) \; \overline{\nu_2(\tilde{b})}$$

where $\tilde{b} \in \tilde{B}_M^{P_1}$ is a real (**i.e.** $\rho(\tilde{b})$ is real) metaframe at $m \in M$, depends only on m and is independent of the particular metaframe at m on which ν_1 and ν_2 are evaluated. In other words, corresponding to each $\nu_1 \in \Gamma(L^{P_1})$ and $\nu_2 \in \Gamma(L^{P_2})$, there is a well defined function $\nu_1 . \bar{\nu}_2 \in C^\infty(M)$.

If $\psi_1 = s_1 . \nu_1 \in W^{P_1} \subset \Gamma(L \otimes L^{P_1})$ and $\psi_2 = s_2 . \nu_2 \in W^{P_2} \subset \Gamma(L \otimes L^{P_2})$ then the function $\psi_1 * \psi_2$ is defined by:

$$\psi_1 * \psi_2 = e^{n\pi i/_4} . (s_1, s_2) . \nu_1 \bar{\nu}_2 \qquad 7.42$$

(The factor $e^{n\pi i/_4}$ ensures that $\psi_1 * \psi_2 = \overline{\psi_2 * \psi_1}$ holds: this can be seen clearly in the second example at the end of this section). For a certain subset of $W^{P_1} \times W^{P_2}$ the complex number:

$$<\psi_1, \psi_2> = \int_M \psi_1 * \psi_2 \ \omega^n \qquad 7.43$$

will be finite and, with luck, will define a unitary transformation:

$$U_{P_1 P_2} : H^{P_1} \to H^{P_2}$$

It should be remarked that there is at least a reasonable hope that the right hand side of eqn. 7.43 will converge on a sufficiently large subset of $W^{P_1} \times W^{P_2}$: ψ_1 and ψ_2 can be chosen to be 'test wave functions', that is so that they vanish outside of $pr_1^{-1}(U_1)$ and $pr_2^{-1}(U_2)$ respectively, where $U_1 \subset X_1 = {}^M/P_1$ and $U_2 \subset X_2 = {}^M/P_2$ are compact and $pr_1 : M \to X_1$ and $pr_2 : M \to X_2$ are the natural projections. In this case, $\psi_1 * \psi_2$ will vanish outside of the set $pr_1^{-1}(U_1) \cap pr_2^{-1}(U_2)$.

In some cases, at least, the BKS transform does give a unitary map $W^{P_1} \to W^{P_2}$ with the required property (see the first example at the end of this section). It also provides a formal solution[29] to the problem of quantizing observables which are not polarization preserving, (at least in the case of real polarizations).

Suppose that (M,ω) is quantizable, that P is a real polarization and that $\phi \in C^{\infty}_{\mathbb{R}}(M)$ is an observable. If ξ_{ϕ} is complete, it will generate a one parameter family of canonical transformations and, for small $t \in \mathbb{R}$, a new polarization P_t can be defined by dragging P a distance t along ξ_{ϕ}.

The observables which can be quantized in this formal sense are those which fulfil the additional requirement that P and P_t are transverse for each small $t \in \mathbb{R}$. One can thus define a family of unitary transformations $W^P \to W^{P_t}$ as follows: let $\psi = s.\nu \in W^P$. Each basis in $B^P(M)$ is dragged by ξ_{ϕ} into a basis in $B^{P_t}(M)$, thus ν is dragged by ξ_{ϕ} into a $^1/_2$-P_t-form $\overset{(t)}{\nu}$. Similarly, by dragging s along η_{ϕ} in L^*, one obtains a new section $\overset{(t)}{s}_t$ which is covariantly constant in the directions in P_t. Combining these two operations, ψ is dragged by ξ_{ϕ} into a section $\overset{(t)}{\psi} = \overset{(t)}{s}.\overset{(t)}{\nu}$ of $L \otimes L^{P_t}$: it is not hard to see that $\overset{(t)}{\psi} \in W^{P_t}$ and that

$$\tilde{\tilde{\delta}}^t_{\phi} : W^P \to W^{P_t} : \psi \longrightarrow \overset{(t)}{\psi}$$

is unitary. On taking the composition of this with the BKS transform $U_{P_t P} : W^{P_t} \to W^P$, one obtains a family of linear transformations:

$$\tilde{\delta}^t_\phi : W^P \to W^P$$

When the BKs transform is well defined, $-(\frac{1}{2\pi i})$ times the generator of this family will be a (hopefully Hermitian) operator $\tilde{\delta}_\phi$ on W^P: this is taken to be the quantum operator corresponding to ϕ.

Unfortunately, this procedure suffers frrom a number of serious drawbacks:

1) It works only for observables generating complete Hamiltonian vector fields.

2) It works only for observables for which P and P_t are transverse for each small t.

3) Even if P and P_t are transverse for every small t, the BKS transform may not be well defined and even if it is, it may not be unitary.

4) In general:

$$-2\pi i.\tilde{\delta}_{[\phi,\psi]} \neq [\tilde{\delta}_\phi,\tilde{\delta}_\psi]$$

5) In general, $\tilde{\delta}^t_\phi$ is not a one parameter group of unitary transformations (this is disastrous in the group theoretic context).

Nevertheless, as the second example illustrates, the procedure yields the physically 'correct' operator in certain familiar classical problems.

Examples:

I) Consider a free particle moving in the Euclidean space $X = \mathbb{R}^n$. The corresponding phase space is $M = \mathbb{R}^{2n}$ with the canonical symplectic form:

$$\omega = dp_a \wedge dq^a \qquad 7.44$$

Here the q^a's are the natural linear coordinates on $\mathbb{R}^n$ and the p_a's are the corresponding components of momentum.

As before, M is quantizable and the line bundle L is the trivial bundle $L = M \times \mathbb{C}$ with the connection form:

$$\alpha = p_a \, dq^a + \frac{1}{2\pi i} \cdot \frac{dz}{z} \quad . \qquad 7.45$$

Sections of L are simply smooth functions on M.

In this case, M admits two naturally defined polarizations

1) The vertical polarization, P_1, spanned by $\{\partial/\partial p_a\}$

2) The horizontal polarization, P_2, spanned by $\{\partial/\partial q^a\}$.

There is only one possible joint metalinear structure for P_1 and P_2 and that is the trivial one: the vector fields $\{\partial/\partial p_a\}$ define a global trivialization of $B^{P_1}(M)$ and so any point of $B^{P_1}(M)$ can be represented by a pair

$$(m, \begin{bmatrix} g & 0 \\ 0 & z \end{bmatrix}) \quad ; \quad \begin{bmatrix} g & 0 \\ 0 & z \end{bmatrix} \in ML(n, \mathbb{C})$$

The $^1/_2$-P_1-form ν_1 defined by:

$$\nu_1(m, \begin{bmatrix} 1 & 0 \\ 0 & 1 \end{bmatrix}) = 1 \quad \forall \quad m \in M \tag{7.46}$$

vanishes nowhere and is constant in the directions in P_1. Hence any wave function $\psi_1 \in W^{P_1}$ can be written:

$$\psi_1 = \phi_1 \cdot \nu_1 \;; \quad \phi_1 \in C^\infty(M) \tag{7.47}$$

where

$$\nabla_\xi \phi_1 = \xi \phi_1 = 0 \quad \forall \quad \xi \in \mathcal{U}_{P_1}(M) \;, \tag{7.48}$$

that is, ϕ_1 is independent of the p_a's.

Similarly, any $\psi_2 \in W^{P_2}$ can be written uniquely in the form:

$$\psi_2 = \phi_2 \cdot \nu_2 \tag{7.49}$$

where ν_2 is the $^1/_2$-P_2-form, given as a function on $\tilde{B}^{P_1}(M)$ by:

$$\nu_2(m, \begin{bmatrix} 1 & 0 \\ 0 & 1 \end{bmatrix}) = 1 \quad \forall \quad m \in M \tag{7.50}$$

and where

$$\nabla_\xi \phi_2 = \xi \phi_2 + 2\pi i(\xi \lrcorner \theta) \cdot \phi_2 = 0 \;\forall\; \xi \in \mathcal{U}_{P_2}(M) \quad . \tag{7.51}$$

Since $\theta = p_a\, dq^a$, ϕ_2 must be of the form

$$\phi_2 = \exp(-\,2\pi i\; p_a\, q^a) \cdot \chi_2(p_a) \qquad 7.52$$

where $\chi_2 \in C^\infty(M)$ is independent of the q^a's. Now

$$\nu_1 \cdot \bar{\nu}_2(m) = 1 \quad \forall \quad m \in M \qquad 7.53$$

so the pairing $W^{P_1} \times W^{P_2} \to C^\infty(M)$ is given by

$$\psi_1 * \psi_2\,(q^a,p_a) = \exp\left(-\,\frac{n\pi i}{4}\right) \cdot \exp(-\,2\pi i . p_a q^a) \cdot \phi_1(q^a) \cdot \chi_2(p_a) \quad 7.54$$

from which it follows that the BKS transform $W^{P_1} \to W^{P_2}$ is given by $\phi_1 \longmapsto \chi_2 = F(\phi_1)$ where F is the ordinary Fourier transform.

II) As a second example, consider the problem of quantizing a free particle moving in a complete oriented Riemannian manifold (X,G). Here the phase space (which is certainly quantizable) is $M = T^*X$, the Hamiltonian is

$$h: M \to \mathbb{R} \;:\; h(q^a,p_a) = {}^1/_2\; G^{ab}\; p_a p_b \qquad 7.55$$

(in local canonical coordinates) and there is a natural polarization P (the vertical polarization).

The problem is to write down the quantum operator on W^P corresponding to the classical observable h.

Now ξ_h is certainly complete but, unfortunately it does not fulfil the second requirement: in general P and P_t are not transverse for any $t \in \mathbb{R}$. In the generic case, for any real t, and for any $x \in X$, there will always be a covector $p \in T_x^* X$ such that $\exp_x(tp)$ is conjugate to x along the geodesic through x with tangent p, so that if $\gamma : t \mapsto \gamma(t)$ is the integral curve of ξ_h through (x,p), then P and P_t will not be transverse at $\gamma(t)$.

Undaunted, I shall ignore this problem and proceed with a formal computation of $\tilde{\delta}_h$.

First, the metalinear structure for P: since X is oriented, it is possible to find, in some neighbourhood U_x of each point $x \in X$, a set of vector fields:

$$\zeta_1 \ldots\ldots \zeta_n$$

which define an oriented orthonormal frame at each point of U_x. The dual bases, which form an orthonormal set of covector fields $\{\alpha^1, \ldots \alpha^n\}$ on U_x, lift to a collection $\{\xi_1 \ldots \xi_n\}$ of vector fields on $pr^{-1}(U) \in M$, given by:

$$\xi_i \lrcorner \omega + pr^* (\alpha^i) = 0$$

(where pr: $T^* X \to X$ is the projection). The ξ_i's are tangent to the fibres in M and so define a local trivialization for $B^P(M)$ in $pr^{-1}(U)$. Each point $b = (m, \eta_1 \ldots \eta_n) \in B^P(pr^{-1}(U))$ is then represented by a pair (m,g) where $m \in M$ and $g \in GL(n,\mathbb{C})$ and

$$\eta_j = (\xi_i)_m \; g_{ij} \qquad 7.56$$

There is an essentially trivial metalinear structure for P in which the two metaframes covering $(m,(\xi_1)_m \ldots (\xi_n)_m)$ at each point $m \in M$ are represented by the $(n + 2)$-tuples $(m,(\xi_1)_m \ldots (\xi_n)_m , 1)$ and $(m,(\xi_1)_m \ldots (\xi_n)_m , -1)$; (they correspond to the pairs:

$$(m, \begin{bmatrix} 1 & 0 \\ 0 & 1 \end{bmatrix}) \quad \text{and} \quad (m, \begin{bmatrix} 1 & 0 \\ 0 & -1 \end{bmatrix})$$

in the local trivialization defined by $\xi_1 \ldots \xi_n$). This is consistent since on the intersection $U_x \cap U_{x'}$ of two of the neighbourhoods in X, the corresponding sets of orthonormal vector fields $\{\zeta_1 \ldots \zeta_n\}$ and $\{\zeta_i' ,\ldots \zeta_j'\}$ are related by:

$$\zeta'_j = \zeta_i \, h_{ij} \qquad 7.57$$

where $h: U_x \cap U_{x'} \to SO(n)$ (that is, for each $y \in U_x \cap U_{x'}$, $h(y)$ is a real orthogonal matrix with unit determinant).

It is thus possible to define a nowhere vanishing section $\nu_o : M \to L^P$ by:

$$\nu_o(m,(\xi_1)_m \ldots (\xi_n)_m, 1) = 1 \text{ in } pr^{-1} (U_x) \qquad 7.58$$

Since ν_o is constant on the fibres in M (in the sense of eqn.

7.27) and since the line bundle L is trivial, each wave function $\psi \in W^P$ can be written in the form:

$$\psi = \phi \cdot \nu_o \qquad 7.59$$

where $\phi \in C^\infty(M)$ is a smooth function independent of momentum. The inner product on W^P is then given by:

$$<\psi_1, \psi_2> = \int_X \phi_1 \cdot \bar{\phi}_2 \; \Omega \qquad 7.60$$

where $\psi_1 = \phi_1 \cdot \nu_o$, $\psi_2 = \phi_2 \cdot \nu_o$ and Ω is the metric volume element.

Now in U_x :

$$pr_*(\mathcal{L}_{\xi_h} \xi_i) = \zeta_i \qquad 7.61$$

Thus for small positive t:

$$(\nu_o \cdot \overline{\nu_o^{(t)}})(m) = t^{n/2} + \text{higher order terms.}$$

Also, if $f: M \times \mathbb{R} \to M$ is the flow generated by ξ_h then:

$$\phi^{(t)}(m) = \phi(f(m,-t)) \exp(-2\pi i \int_o^t (A - h)) \qquad 7.62$$

where A (= 2h in this case) is the action function and the integral is taken along the orbit $t \longmapsto f(m,t)$. Thus, if $\psi = \phi \cdot \nu_o$, then:

$$\langle\psi^{(t)},\psi\rangle = e^{n\pi i/4}\, t^{n/2} \int_M \bar{\phi}(m).\phi(f(m,-t)).\exp(-2\pi i\int_0^t (A - h))\omega^n \qquad 7.63$$

if ϕ is taken to be a test function with support contained in some U_x, then this integral can be simplified by introducing (non-canonical) coordinates into $pr^{-1}(U_x)$. These are defined by first choosing arbitrary coordinates q^a in U_x. The vector fields $\zeta_1, \ldots, \zeta_n$ are then given by:

$$\zeta_i = \zeta_i^a \frac{\partial}{\partial q^a} \qquad 7.64$$

and the dual bases by:

$$\alpha^i = \alpha^i_a \, dq^a$$

The coordinates of $(y,p) \in pr^{-1}(U)$ are taken to be (q^a, p_a) where q^a are the coordinates of y and

$$p = p_i \, \zeta^i_a \, dq^a = p_i \, \alpha^i \qquad 7.65$$

In these coordinates, the volume element ω^n is given by:

$$\omega^n = g^{\frac{1}{2}} \, d^n q \, d^n p \qquad 7.66$$

where $g = \det(g_{ab})$ and the integral (eqn. 7.63) becomes:

$$\langle \psi^{(t)}, \psi \rangle = e^{n\pi i/4}\, t^{n/2} \iint e^{-\pi i t(p_1{}^2 + \ldots + p_n^2)}\, \bar{\phi}(q)\, \phi(\exp_q(-tp)) \times g^{\frac{1}{2}}\, d^n q\, d^n p \qquad 7.67$$

$$= e^{n\pi i/4}\, t^{-n/2} \iint e^{-\pi i(p_1{}^2 + \ldots + p_n^2)/t} \cdot \bar{\phi}(q) \cdot \phi(\exp_q(-p)) \times g^{\frac{1}{2}}\, d^n q\, d^n p \qquad 7.68$$

by a change of variable.

To find the quantum operator $\tilde{\delta}_h$, one could compute from this the BKS transform $\widetilde{\psi^{(t)}}$ of $\psi^{(t)}$ for each t and then take the generator of the one parameter family of unitary maps $\psi \longmapsto \widetilde{\psi^{(t)}}$. However, at least formally, this can be achieved more simply by differentiating eqn 7.68 with respect to t and taking the limit as $t \to 0$, using the lemma:

<u>Lemma</u>: If $f(u,t) = t^{-\frac{1}{2}} \exp(-\frac{\pi i u^2}{t})$ then

1) $f(u,t) \to 2e^{-\pi i/4}\, \delta(u)$ as $t \to 0$ (as a generalized function)

2) For each $t \in \mathbb{R}$, $\dfrac{\partial^2 f}{\partial u^2} = 4\pi i \dfrac{\partial f}{\partial t}$.

<u>Proof</u>: The first part follows by applying the general test[30] for δ convergent sequences, using the classical integrals of Fresnel. The second part is trivial.

Thus differentiating eqn. 7.68 and taking the limit as $t \to 0$:

$$- 2\pi i \; \langle \tilde{\delta}_h \psi, \psi \rangle = \frac{1}{2\pi i} \iint \Big(\sum_{i=1}^{n} \delta(p_1) \ldots \delta''(p_i) \ldots \delta(p_n) \Big)$$

$$\bar{\phi}(q) \, \phi(\exp_q(-p)) \, g^{\frac{1}{2}} \, d^n q \, d^n p \; .$$

Now, from elementary differential geometry:

$$\frac{\partial^2}{\partial p_i^2} (\phi(\exp_q(-p))) \Big|_{p=o} = - \zeta_i^a \, \zeta_i^b \, \nabla_a \nabla_b \, \phi$$

where ∇ is the metric connection on (X,g). Integrating over p therefore:

$$- 2\pi i . \; \tilde{\delta}_h \, \psi = - \frac{1}{2\pi i} \quad (\Delta \, \phi) \, . \, \nu_o$$

where Δ is the Laplace-Beltrami operator.

8. Invariance Groups

Historically, the theory of group representations played a central role in the development of quantum mechanics. This was a reflection of the axiom that a symmetry group (such as the Poincaré or Galilei group) which appears at the classical level in some physical system must also act as a symmetry group in the underlying quantum system: in any quantization scheme, a classical group of canonical transformations must emerge as an invariance group in the quantum phase space[31].

This thinking has led to an approach to quantum theory which bypasses Hamiltonian mechanics altogether, at least at the kinematical level. In order to find a quantum model of, for example, an elementary relativistic particle, it is unnecessary (so the argument goes) to quantize the corresponding classical system. All one need do is to find a suitable irreducible unitary representation of the Poincaré group.

In the past, however, the detailed connection between this and the more conventional methods of quantum theory has often been obscure. First, because there was usually no obvious natural relationship between a quantum Hilbert space built up, for example, out of the wave functions on the classical configuration space and one constructed abstractly using group theoretic techniques (though, of course, there are general theorems which state that they must be the same) and secondly because it was not often clear what the corresponding classical system was in any case. Indeed, for a long time it was thought that certain elementary particles (such as those with spin) had no analogues in classical mechanics.

By exploiting the geometrical techniques outlined in the preceding sections, Kostant and Souriau have gone a long way towards clearing up this obscurity. The cornerstone of their work is a method (due originally to Kirillov) for finding all the elementary classical systems which admit a particular group as an invariance group, that is all the classical phase spaces which admit the group as a transitive group of canonical transformations[32]. The corresponding elementary quantum systems can then be found by geometric quantization: in other words, one and the same technique is used for finding an irreducible representation of the group as for quantizing the corresponding elementary classical system.

Their idea is to build up the elementary classical systems of a given invariance group G from the orbits of G in $\mathcal{G}^*$ (the dual of its Lie algebra). Each of these is naturally a homogeneous symplectic G-manifold, that is a symplectic manifold on which G acts as an invariance group.

In detail, it works like this: let G be a connected simply connected Lie group (discrete symmetries will be ignored) and let $\mathcal{G}$ be the Lie algebra of G: $\mathcal{G}$ can be identified with the tangent space to G at the identity, $e \in G$. Each $g \in G$ defines a diffeomorphism τ_g of G which preserves the identity:

$$\tau_g : x \longmapsto gxg^{-1} \quad ; \quad x \in G$$

The derivative of τ_g at e is therefore a linear transformation of $\mathcal{G}$, denoted Ad_g; the map $g \longmapsto Ad_g$ is called the <u>adjoint representation</u> of G; it satisfies $Ad_{gg'} = Ad_g Ad_{g'}$ for all $g, g' \in G$.

Thus G acts on $\mathcal{G}$ as a group of linear transformations. This action induces a second action Ad' on the dual space $\mathcal{G}^*$, called the <u>coadjoint representation</u>. Explicitly:

$$(Ad'_g\, f)\,(X) = f(Ad_{g^{-1}}\, X) \quad ; \quad f \in \mathcal{G}^*, \; X \in \mathcal{G}\,, \; g \in G. \tag{8.1}$$

For simplicity, $Ad'_g \cdot f$ will be written $g \cdot f$.

Suppose now that

$$M_{f_o} = \{g \cdot f_o \mid g \in G\} \quad ; \quad f_o \in \mathcal{G}^* \tag{8.2}$$

is an orbit in $\mathcal{G}^*$ and that $f \in M_{f_o}$. Each $X \in \mathcal{G}$ generates a one parameter subgroup of G and hence defines a flow on M_{f_o} ; let ξ_X be the tangent vector field to this flow. The map:

$$\mathcal{G} \to T_f M_{f_o} : X \longmapsto X_f = (\xi_X)_f$$

is linear and surjective (since the action of G on M_{f_o} is transitive). Also, if $X, X' \in \mathcal{G}$, then $X_f = X'_f$ if, and only if,

$$X = X' + Z$$

where $Z \in G$ is such that $\exp(tZ)$ leaves f invariant for each $t \in \mathbb{R}$. From the form of the coadjoint representation, this is equivalent to:

$$f([Z,X]) = 0 \qquad \forall \; X \in G \tag{8.3}$$

Thus the quantity ω_f, given by:

$$\omega_f(X_f, Y_f) = f([X,Y]) \; ; \qquad X, Y \in G \tag{8.4}$$

is a well defined skew symmetric bilinear form on $T_f M_{f_o}$; it is also non-degenerate since, if $X \in G$ then:

$$\omega_f(X_f, Y_f) = 0 \qquad \forall \; Y \in G \tag{8.5}$$

if, and only if, each $\exp(tX)$ leaves f invariant, that is, if and only if, $X_f = 0$.

As f varies, ω_f defines a non-degenerate 2-form $\omega \in \Omega^2(M_{f_o})$; to show that ω is, in fact, a symplectic structure, it is only necessary to prove that $d\omega = 0$. Not surprisingly, this follows from the Jacobi identity in G. In fact, if $X, Y \in G$ then:

$$[\xi_X, \xi_Y]_f = [X,Y]_f \tag{8.6}$$

Hence if X, Y, Z $\in$ G then:

$$(d\omega(\xi_X,\xi_Y,\xi_Z))_f = \sum_{\text{cyclic}} X_f(\omega(\xi_Y,\xi_Z)) - f([[X,Y],Z]) \qquad 8.7$$

The second term is zero by the Jacobi identity in G. The first term can be computed using the fact that, for fixed X $\in$ G, the rate of change of f(X) along ξ_Z, where Z $\in$ G, is f([Z,X]) since, for small $t \in \mathbb{R}$:

$$((\exp(-tZ)).f)(X) = f(\exp(tZ).X)$$

$$= f(X + t[Z,X]) + O(t^2) \qquad 8.8$$

Thus:

$$X_f(\omega(\xi_Y,\xi_Z)) = f([X,[Y,Z]] \qquad 8.9$$

and so the first term also vanishes by the Jacobi identity.

Finally ω is invariant under the action of G on M_{f_0} since, for any X, Y, Z $\in$ G :

$$((\mathcal{L}_{\xi_X}\omega)(\xi_Y,\xi_Z))_f$$

$$= X_f(\omega(\xi_Y,\xi_Z)) - \omega([\xi_X,\xi_Y],\xi_Z)_f - \omega(\xi_Y,[\xi_X,\xi_Z])_f$$

$$= f([X, [Y,Z]]) + [Z,[X,Y]] + [Y; [Z,X]])$$

$$= 0 \qquad 8.10$$

Thus each orbit in G^* has the structure of an (albeit abstract) classical phase space on which G acts as a transitive invariance group. The importance of this result is that essentially <u>all</u> classical phase spaces which admit G as a transitive invariance group (that is all homogeneous symplectic G manifolds) arise in this way.

To understand the qualification 'essentially' it is necessary to return to the problem, which was first raised in §4, of associating a classical observable with each generator of a given symmetry group. First, some notation. A symplectic manifold (M,ω) is called a <u>Hamiltonian G-space</u> for a Lie group G if there is given a Lie algebra homomorphism:

$$\lambda : \mathcal{G} \to C^{\infty}_{\mathbb{R}}(M) : X \in \phi_X$$

from the Lie algebra of G into the space of real functions (observables) on M such that[33]:

1) Each Hamiltonian vector field $\xi_X = \xi_{\phi_X}$ is complete
2) Any two points $m_1, m_2 \in M$ can be joined by an integral curve of ξ_X for some $X \in \mathcal{G}$.

Every Hamiltonian G-space is also a homogeneous symplectic G-manifold:[34] the action of each $g \in G$ is defined by integrating the (complete) Hamiltonian vector field of the corresponding generator in $\mathcal{G}$. Moreover, and this is the important point, every Hamiltonian G-space is a covering space of an orbit in[34] $\mathcal{G}^*$. The proof of this is almost trivial: if (M,ω) is a Hamiltonian G-space then the map

$$M \to G^* : m \mapsto f_m$$

defined by:

$$f_m(X) = \phi_X(m) \; ; \quad m \in M, \quad X \in G \qquad 8.11$$

commutes with the actions of G on M and G^* and so maps M onto an orbit in G^*. It is not hard to see that it is, in fact, a covering map.

Suppose now that there is given a classical system with a phase space (M,ω) and a transitive invariance group G. If it is possible to find a map $\lambda : G \to C^\infty_{\mathbb{R}}(M)$ which generates the action of G and which makes (M,ω) into a Hamiltonian G-space, then (M,ω) can be identified with a covering space of an orbit in G^* (in fact, if M is simply connected then it must actually be an orbit in G^*); (M,ω) can then be classified purely in terms of the structure of G.

For λ to exist, two conditions must be satisfied (it is these that are embodied in the qualification 'essentially'). First, each generator $X \in G$ of G defines a one parameter group of canonical transformations of M, and hence a locally Hamiltonian vector field ξ_X : for λ to exist, each ξ_X must in fact be globally Hamiltonian. This will be so if M is simply connected or (as in the case of SO(3)) if $G = [G,G]$ (for example, if G is semi-simple). Secondly, even if ϕ_X can be found for each $X \in G$ individually, it will not necessarily follow that λ preserves brackets, that is that:

$$\phi_{[X,Y]} = [\phi_X, \phi_Y] \qquad \forall \; X,Y \in G \quad . \qquad 8.12$$

The condition that each ϕ_X can be chosen so that this is true involves the cohomology of G : it is explained in appendix C. However, it is always true (provided the first condition is satisfied) that λ can be found for some central extension of G (again, see appendix C).

At the purely classical level, therefore, this construction provides an elegant classification scheme for the elementary systems with a given invariance group. At the quantum level it assumes a more important role. For suppose that (M,ω) is a quantizable symplectic manifold and that G is a transitive invariance group. The symmetry axiom requires that G should act as a symmetry group on the phase space of the underlying quantum system; also, according to the argument given in §5, this action should be irreducible. If (M,ω) is, in fact, a Hamiltonian G-space, and G is simply connected[35], then the first part, at least, will automatically be achieved by geometric quantization: each generator $X \in G$ is associated with a classical observable ϕ_X and hence with a vector field $\eta_X = \eta_{\phi_X}$ on the prequantization line bundle, L. This vector field will be complete (since ξ_X is complete) and will generate a one parameter family of unitary transformations of $\Gamma(L)$. Thus $\exp(X)$ and hence G has a natural action on $\Gamma(L)$. An irreducible action can usually be achieved by choosing a G-invariant polarization of M.

The condition that (M,ω) should be a Hamiltonian G-space for its invariance group thus plays two roles. First, it allows (M,ω) to be identified with a (covering space of) an orbit in G^* and secondly it ensures that the quantization procedure is natural, that is that G is also a symmetry group at the quantum level.

9. Examples

This final section contains three examples which illustrate both the usefulness of the theory and its limitations.

I. Free particles and the WKB approximation:

Consider a classical conservative system with configuration space X (a smooth n-dimensional manifold). The system has as phase space, the cotangent bundle $M = T^*X$, and it can be quantized using the vertical polarization by the method described in §7: in local canonical coordinates $\{q^a, p_a\}$ (defined by coordinates $\{q^a\}$ on X) the wave functions have the form:

$$\psi = \phi(q^a)\,(d^n q)^{\frac{1}{2}} \qquad 9.1$$

However, the vertical polarization P is not the only possible choice. Other polarizations can be constructed as follows: let $f\colon X \to \mathbb{R}$ be any smooth function. At least locally, the image Λ of the map

$$\phi : X \to T^*X : x \longmapsto (x, (df)_x) \qquad 9.2$$

is an n-dimensional surface in T^*X. Furthermore,

$$\phi^*(\theta) = df \qquad 9.3$$

(where θ is the canonical 1-form) since, for any tangent vector $\zeta \in T_x X$ at $x \in X$:

$$\zeta \lrcorner (\phi^* \theta) = (\phi_* \zeta) \lrcorner \theta = (pr_* \phi_* \zeta) \lrcorner df = \zeta \lrcorner df \qquad 9.4$$

(here $pr : T^* X \to X$ is the projection map). Thus:

$$\phi^* \omega = d(\phi^* \theta) = 0 \qquad . \qquad 9.5$$

In other words, the (complexified) tangent space to Λ is a maximally isotropic subspace of $(T_m M)^{\mathbb{C}}$ at each $m \in \Lambda$.

It follows that any n-parameter family of smooth functions

$$S_k : X \to \mathbb{R} : x \longmapsto S(k,x); \; k = (k_1 \dots k_n) \in \mathbb{R}^n$$

with the property that the images of the maps:

$$\phi_k : X \to T^* X : x \longmapsto (x, (dS_k)_x)$$

fill out an open set in $T^* X$, will (at least locally) define a polarization Q of $T^* X$; for each $m = (x,p) \in T^* X$, Q_m is the (complexified) tangent space to the $\phi_k(X)$ through m. This polarization (again, only locally) will be transverse to P.

If, additionally, the function S is chosen so that for each $k \in \mathbb{R}^n$:

$$h \circ \phi_k = \text{const.} \qquad 9.6$$

where $h: T^*X \to \mathbb{R}$ then ξ_h will be tangent to the leaves of Q (the surfaces $\phi_k(X)$) and h can be quantized directly. Now eqn. 9.6 can be rewritten in local coordinates:

$$h(q^a, \frac{\partial S_k}{\partial q^a}) = \text{const.} \qquad 9.7$$

which is nothing more than the time independent form of the Hamilton-Jacobi equation: the functions S_k must form a complete integral of the Hamilton-Jacobi equation. Complete integrals always exist locally, though **there** may not be any well behaved global solutions. However, one can still proceed with the quantization by restricting attention to a small portion of phase space in which there is precisely one surface $\phi_k(X)$ through each point (however, see note 37).

(It sometimes happens, that, even though the functions S_k are well defined only locally, the local polarization Q can be extended to a global polarization. For example, a free particle moving in $X = \mathbb{R}^2 - \{0\}$ has Hamiltonian

$$h = \frac{1}{2} (p_r^2 + p_\theta^2/r^2) \qquad 9.8$$

(in polar coordinates). A complete integral of the corresponding Hamilton-Jacobi equation is:

$$S_k = \Theta(\theta) + R(r) \qquad 9.9$$

where $k = (h,\ell) \in \mathbb{R}^2$ and:

$$(\Theta')^2 = \ell^2; \quad (R')^2 = 2h - \ell^2/r^2. \qquad 9.10$$

The 2-surfaces in T^*X defined by:

$$p_\theta{}^2 = \ell^2; \quad p_r{}^2 = 2h - \ell^2/r^2 \qquad 9.11$$

are the leaves of a well behaved polarization Q, but the functions S_k are singular and are either double valued or not defined at all in the large. The singularities correspond to the points where Q is not transverse to P.)

Working locally and choosing the prequantization line bundle as before, the Q wave functions have the form:

$$\psi(q^a,k_a) = \phi(k_a)e^{-2\pi i.S(q^a,k_a)}.(d^nk)^{\frac{1}{2}} \qquad 9.12$$

where $\{q^a,k_a\}$ are used as local non-canonical coordinates on M.

Since ξ_h is tangent to the leaves of Q:

$$\tilde{\delta}_h \psi = h\psi \qquad 9.13$$

Now it is always possible to choose one of the parameters k_a (k_n, say) to be h. When this is done, the distributional wave function

$$\psi(q^a, k_a) = (k_a - k_a^{(o)})\, e^{-2\pi i\, S(k_a^{(o)}, q^a)} (d^n k)^{\frac{1}{2}} \qquad 9.14$$

will be an eigenstate of $\tilde{\delta}_h$, with eigenvalue $h^{(o)} = k_n^{(o)}$ (here the $k_a^{(o)}$'s are constants). A calculation similar to that used in §7 (example I) shows that, under the BKS-transform, ψ is mapped onto the P-wave function

$$\chi = e^{-2i\pi S(k_a^{(o)}, q^a)} \cdot \left|\det\left(\frac{\partial^2 S}{\partial q^a\, \partial k_b}\right)\right|^{\frac{1}{2}} (d^n q)^{\frac{1}{2}} \quad . \qquad 9.15$$

If $h = \frac{1}{2} g^{ab} p_a p_b$ is the Hamiltonian of a free particle, χ is not an eigenfunction of the corresponding Laplace-Beltrami operator (acting on $\frac{1}{2}$-forms). It is a second order WKB-approximation to an eigenfunction. (To be precise, if one replaces 2π by $1/\hbar$ in eqn. 9.15 and if one ignores terms of order h^2, then χ will satisfy

$$\frac{1}{2}\hbar^2 \Delta \chi + h^{(o)}\chi = 0 \qquad 9.16$$

where Δ is the Laplace-Beltrami operator.)

One can turn this around and ask: given an observable k, for example, a homogeneous polynomial in the momenta, what differential operators δ_k on W^P have eigenfunctions with WKB approximations of the form 9.15?

Locally, such an observable has the form

$$k(q,p) = K^{ab\ldots d}\, p_a p_b \cdots p_d \qquad 9.17$$

where the $K^{ab\ldots d}$'s are the components of a symmetric tensor of type $\binom{N}{0}$ (say); a short calculation reveals that δ_k must have the form

$$\delta_k(\psi|d^nq|^{\frac{1}{2}}) = (2\pi i)^{-N}\, K^{ab\ldots d}\, \partial_a \partial_b \cdots \partial_d \psi$$

$$+ \tfrac{1}{2}N(\partial_a\, K^{ab\ldots d})\, \partial_b \cdots \partial_d \psi\, |d^nq|^{\frac{1}{2}} + \delta'_k(\psi|d^nq|^{\frac{1}{2}}) \qquad 9.18$$

where δ'_k is a differential operator of degree N-2 (again, strictly, one should replace 2π by $1/\hbar$ and expand in powers of $\hbar$, keeping only the first two terms).

Thus, if all choices of S_k are treated equally, the quantization of k as an operator on W^P is ambiguous: the answer is only determined up to the addition of an operator of degree N-2. This same result can be obtained in a slightly different way: an observable k of the form 9.17 can be written as the sum of products of observables of the form

$$z(q, p) = \zeta^a p_a \qquad 9.19$$

where $\zeta = \zeta^a \partial_a$ is a real vector field on X, and can thus be quantized by replacing each z by $\tilde{\delta}_z$ (which is well defined since ξ_z preserves P). More formally, k can be quantized by identification with an element of the universal enveloping algebra U of the Lie subalgebra of observables which preserve P. Unfortunately, the identification of k with an element of U is not unique: there are many different ways of writing k as a sum of products of observables of the form 9.19 and, in general, these will lead to different quantizations of k. However the result of this procedure is that k is quantized as a differential operator on W^P of the form 9.18: <u>the ambiguity is the same.</u>

The conclusion to be drawn from this is that in a general mechanical system, the quantization procedure <u>depends critically on the choice of polarization</u>. Quantization in an arbitrary polarization will result in only a semi-classical approximation to the 'true' quantum theory.

In certain situations this is not a serious difficulty. For example, for scalar particles and for spinning particles in Minkowski space (see below) there is a naturally defined polarization. For more complex systems (such as spinning particles in curved space[36]), there is no such polarization and no obvious natural way of constructing the quantum theory from the classical data alone[37]: one must be content with a WKB approximation.

What happens when one tries to extend the ideas introduced above to a <u>global</u> treatment of the WKB approximation? The canonical 2-form on T^*X is exact in the large, so the sections of the prequantization line bundle can still be represented by complex valued functions on T^*X. Further, near a point where its leaves are transverse to the fibres of T^*X, any real polarization Q tangent to ξ_h can be represented as before by a complete integral of the Hamilton-Jacobi equation. (This follows by reversing the reasoning which lead to eqn. 9.5.) The trouble comes at points where Q is not transverse to P: these points correspond to the caustics in the congruences classical trajectories of ξ_h which lie in the leaves of Q. Though nothing goes wrong with the Q-quantization at these points, the BKS-transform becomes singular: the expression on the right hand side of eqn. 9.15 blows up. The idea which rescues the scheme is to interpret this expression as a distribution which is well defined even at the non-transverse points. I will not give the details here; I will merely sketch the essential role which the ½-forms play in making this idea precise.

First, returning to the local treatment, suppose that there is a third polarization $\tilde{Q}$ generated by an n-parameter family of functions $\tilde{S}_{\tilde{k}}$ (not necessarily satisfying the Hamilton-Jacobi equation). Then the BKS-transform of the $\tilde{Q}$-wave function

$$\tilde{\psi}(q^a,\tilde{k}_a) = \delta(\tilde{k}_a-\tilde{k}_a^{(o)})e^{-2\pi i\tilde{S}(\tilde{k}_a^{(o)},q^a)}(d^n\tilde{k})^{\frac{1}{2}} \qquad 9.20$$

is the P-wave function

$$\tilde{\chi} = e^{-2\pi i S(\tilde{k}_a^{(o)},q^a)}\left|\det\left(\frac{\partial^2 S}{\partial q^a\partial\tilde{k}_b}\right)\right|^{\frac{1}{2}}(d^n q)^{\frac{1}{2}} \qquad 9.21$$

Using the method of stationary phase, the inner product in W^P of $\tilde{\chi}$ and χ is, at least in the WKB approximation (where one assumes that the phase factors in eqns. 9.15 and 9.21 are rapidly varying),

$$\langle \chi,\tilde{\chi}\rangle = e^{i\pi \operatorname{sign}(A)/4}\, e^{2\pi i(s(q^a)-\tilde{s}(q^a))}\, \Delta_S^{\frac{1}{2}}\, \Delta_{\tilde{S}}^{\frac{1}{2}}\, |\det A|^{\frac{1}{2}} \qquad 9.22$$

where:

(a) $s(q^a) = S(k_a^{(o)},q^a)$ and $\tilde{s}(q^a) = S(\tilde{k}_a^{(o)},q^a)$

(b) the q^a's are the coordinates of the point where $\frac{\partial s}{\partial q^a} = \frac{\partial \tilde{s}}{\partial q^a}$ (geometrically, this is the point of intersection of the $k_a^{(o)}$ leaf of Q with the $\tilde{k}_a^{(o)}$ leaf of $\tilde{Q}$; of course, if there is more than one point of intersection there will be other contributions to eqn. 9.22).

(c) $\Delta_S = |\det \frac{\partial^2 S}{\partial q^a \partial k_b}|^{\frac{1}{2}}$ and $\Delta_{\tilde{S}} = |\det \frac{\partial \tilde{S}}{\partial q^a \partial \tilde{k}_b}|^{\frac{1}{2}}$

(d) A is the matrix $\frac{\partial^2 S}{\partial q^a \partial k_b} - \frac{\partial^2 \tilde{S}}{\partial q^a \partial \tilde{k}_b}$; sign (A) is the signature of A.

One might hope to obtain this inner product directly by pairing ψ and $\tilde{\psi}$ as elements of the wave function spaces W^Q and $W^{\tilde{Q}}$. In fact, this gives the same expression as on the right hand side of eqn. 9.22, but <u>without</u> the factor $e^{i\pi \operatorname{sign}(A)/4}$. To obtain this factor, more care must be taken in assigning metalinear structures to Q and $\tilde{Q}$, and in defining the pairings between P,Q and $\tilde{Q}$. The "joint metalinear structure" introduced in §7 only works for a pair of transverse polarizations.

In the present case, it can only be used locally, that is, in an open submanifold of M in which Q and $\tilde{Q}$ are transverse to each other and to P. Instead of using this naive approach, one must find a way of assigning metalinear structures to <u>all</u> real polarizations simultaneously. This is done by introducing a <u>metaplectic structure</u> on M, as follows:

The bundle of (real) symplectic bases over M has structure group $SP(n,\mathbb{R})$ (the group of linear transformations of $\mathbb{R}^{2n}$ which preserve the 2-form $\begin{bmatrix} 1 & 0 \\ 0 & -1 \end{bmatrix}$). Like the linear group $GL(n,\mathbb{C})$, this group has a double covering $MP(n,\mathbb{R})$ (the <u>metaplectic group</u>), so that, just as before, one can introduce the idea of a <u>metaplectic frame bundle</u> and the idea of a <u>metaplectic frame</u> at each $m \in M$: each metaplectic frame corresponds to two symplectic frames at m. Further, a choice of metaplectic structure defines a unique metalinear structure for each polarization. In particular, if X is oriented, there is a trivial metaplectic structure and the use of the corresponding metalinear structures on Q and $\tilde{Q}$ gives the correct $\exp(i\pi \,\mathrm{sign}(A)/4)$ factor in the pairing of Q and $\tilde{Q}$: the pairing between Q and $\tilde{Q}$ is defined as in §7 except that the product at $m \in M$ of a $\frac{1}{2}$-Q-form with a $\frac{1}{2}$-$\tilde{Q}$-form is computed by introducing metalinear frames for Q and $\tilde{Q}$ at m which together form a metaplectic frame. The point of all this is that it is now possible to define the BKS transform $\chi \in W^P$ of the W^Q wave function ψ (defined by eqn. 9.14) even if the $k_a^{(o)}$-leaf of Q is not everywhere transverse to the fibres in T^*X: the value of χ at $m \in M$ is given by first pairing ψ with the wave functions of some local polarization $\tilde{Q}$ (defined in some neighbourhood of m) transverse to P and Q and then identifying the elements of $W^{\tilde{Q}}$ with W^{P} wave functions as in eqn. 9.21: thus χ is defined as a distribution near m by testing it against W^{P} wave functions of the form 9.21. The result is independent of the choice of $\tilde{Q}$, at least in the stationary phase approximation. (Of course, this is only an outline: to do this in detail one must take much more care over the domains of definition of the various wave functions and so on.)

The only practical difficulty is that it is hard to find a simple explicit form for $MP(n,\mathbb{R})$. This problem can be avoided by introducing an almost-Kähler structure J on M and restricting attention to symplectic frames $\{\xi_1, \ldots \xi_n, \zeta_1, \ldots \zeta_n\}$ which satisfy

$$J(\xi_i) = \zeta_i. \qquad 9.23$$

These will be called <u>unitary frames</u>. The structure group of the bundle of unitary frames is isomorphic to the unitary group U(n). It is the subgroup of $SP(n, \mathbb{R})$ of matrices of the form

$$\begin{bmatrix} A & B \\ -B & A \end{bmatrix}$$

where A and B are real $n \times n$ matrices and $A + iB \in U(n)$. The unitary group is, in fact, the maximal compact subgroup of $SP(n,\mathbb{R})$. It is easy to give an explicit form for the corresponding subgroup of $MP(n,\mathbb{R})$: it is simply the double cover of U(n), that is, the group of $(2n + 1) \times (2n + 1)$ matrices of the form

$$\begin{bmatrix} A & B & 0 \\ -B & A & 0 \\ 0 & 0 & Z \end{bmatrix}$$

where $Z^2 = \det(A + iB)$. This double covering is all that is needed, though, unfortunately, there is now the additional complication of having to decide how the results depend on the choice of J.

To summarize, the introduction of metaplectic structures and $\frac{1}{2}$-forms allows one to carry the WKB approximation through the caustics in the classical trajectories in an entirely natural way. This treatment is equivalent to Maslov's. It automatically incorporates the correct phase jump at the caustics and it leads to Maslov's modification of the Bohr-Sommerfeld rule (see notes 19 and 23).

II The harmonic oscillator:

The second example is the quantization of the n-dimensional harmonic oscillator. Here the phase space is $M = \mathbb{R}^{2n}$ with coordinates $\{q^a, p_a\}$. The symplectic 2-form is:

$$\omega = dp_a \wedge dq^a \qquad 9.24$$

and the Hamiltonian is:

$$h = \frac{1}{2}(p^2 + w^2 q^2); \quad p^2 = \delta^{ab}\, p_a\, p_b, \quad q^2 = \delta_{ab}\, q^a\, q^b \qquad 9.25$$

The vertical polarization P, spanned at each point by $\{\partial/\partial p_a\}$ is not invariant under the action of ξ_h. However M can be given the structure of an n-dimensional complex vector space by introducing the coordinates:

$$z^a = p_a - iwq^a; \quad \bar{z}_a = p_a + iwq^a \qquad 9.26$$

The antiholomorphic polarization $\tilde{P}$ (spanned at each point by the vectors $\{\partial/\partial\bar{z}_a\}$) is then Lie propagated by ξ_h and the quantization procedure can be applied directly[38].

In the new coordinates:

$$\omega = d\theta = \frac{1}{2iw}(dz^a \wedge d\bar{z}_a) \quad 9.27$$

where

$$\theta = \frac{1}{4iw}(z^a\, d\bar{z}_a - \bar{z}_a\, dz^a) \quad 9.28$$

(note that θ is real). Since ω is exact, the prequantization line bundle L can be represented as:

$$L = M \times \mathbb{C} \quad 9.29$$

with the connection form:

$$\alpha = \theta + \frac{1}{2\pi i}\left(\frac{dz}{z}\right) \quad 9.30$$

(using $\{z_1, \ldots, z_n, z\}$ as coordinates on $M \times \mathbb{C}$). Also, the Hamiltonian can be expressed in the form:

$$h = \frac{1}{2} z^a \bar{z}_a \quad 9.31$$

so that

$$\xi_h = -iw\left(z^a \frac{\partial}{\partial z^a} - \bar{z}_a \frac{\partial}{\partial \bar{z}_a}\right) \quad 9.32$$

This vector field has integral curves:

$$t \longmapsto z^a(t) = e^{-iwt} \, . \, z^a(0) \qquad 9.33$$

and its action function is given by:

$$A = \xi_h \lrcorner \, \theta = h \qquad 9.34$$

Polarized sections $m \longrightarrow (m, \phi(m))$ of L are given by functions $\phi: M \to \mathbb{C}$ of the form:

$$\phi : z^a \longmapsto f(z^a) \, e^{-(\pi/_{2w}) \, z^a \, \bar{z}_a} \qquad 9.35$$

where $f : M \to \mathbb{C}$ is holomorphic. Thus the $\tilde{\text{P}}$-wave functions ψ can be written:

$$\psi(z^a, \bar{z}_a) = \phi(z^a, \bar{z}_a) \, . \, \nu^{\frac{1}{2}} = f(z^a).e^{-(\pi/_{2w}) \, z^a \bar{z}_a} \, . \, \nu^{\frac{1}{2}} \qquad 9.36$$

where ν is the holomorphic n-form:

$$\nu = \epsilon_{abc} \ldots dz^a \wedge dz^b \wedge dz^c \wedge \ldots \qquad 9.37$$

(ν, and hence also $\nu^{\frac{1}{2}}$, is Lie propagated by the vector fields $\partial/\partial\bar{z}_a$).

On applying the formula eqn. 7.34, one obtains the inner product of the two wave functions:

$$\psi = f \, . \, e^{-(\pi/_{2w}) \, z^a \, \bar{z}_a} \, . \, \nu^{\frac{1}{2}} \qquad 9.38$$

$$\chi = g \cdot e^{-(\pi/_{2w})\, z^a \bar{z}_a} \cdot \nu^{\frac{1}{2}} \qquad 9.39$$

in the form:

$$<\psi,\chi> = \int_M f\bar{g}\, e^{-(\pi/_w)\, z^a \bar{z}_a} \cdot \omega^n \qquad 9.40$$

which (modulo minor conventional differences) is the same as the expression given by Bargmann[38].

To determine the permitted energy levels, it is only necessary to solve for h_o the eigenvalue equation:

$$- 2\pi i\, \tilde{\delta}_h\, \psi = [(\nabla_{\xi_h} - 2\pi i\, h)\phi]\nu^{\frac{1}{2}} + (\mathcal{L}_{\xi_h} \nu^{\frac{1}{2}}) \cdot \phi$$

$$= - 2\pi i . h_o\, \phi\, \nu^{\frac{1}{2}} \qquad 9.41$$

where ψ is as in eqn. 9.28. Now:

$$\mathcal{L}_{\xi_h} \nu = - iwn \cdot \nu \qquad 9.42$$

so that

$$\mathcal{L}_{\xi_h} \nu^{\frac{1}{2}} = - \frac{iwn}{2} \cdot \nu^{\frac{1}{2}} \qquad 9.43$$

Thus the eigenvalue equation reduces to:

$$\xi_h\, f - \frac{iwn}{2} \cdot f = - 2\pi i\, h_o \cdot f \qquad 9.44$$

However, the orbits of ξ_h have period $2\pi/w$ so that, for solutions of eqn 9.36 to exist

$$N = \frac{2\pi}{w}\left(h_o - \frac{w}{2\pi} \cdot \frac{n}{2}\right) \qquad 9.45$$

must be an integer. Eqn 9.36 then becomes:

$$z^a \frac{\partial f}{\partial z^a} = Nf \qquad 9.46$$

This has nonsingular solutions only for $N \geq 0$, in which case f is a homogeneous polynomial of degree N. Thus the energy levels are given by:

$$h_o = \frac{w}{2\pi}\left(n + {}^{n}/_{2}\right) \quad ; \qquad N = 0, 1, 2 \ldots \qquad 9.47$$

with multiplicities $({}^{1}/_{N!})\ n(n+1) \ldots (n+N-1)$ (the number of homogeneous polynomials in n variables of degree N). This agrees with the usual Schrödinger treatment.

The fact that the Kostant-Souriau procedure gives the correct answer seems to be a coincidence, closely related to the fact that the Bohr-Sommerfeld quantization (with the Maslov correction[39]) also gives the correct energy levels for the harmonic oscillator.

III The Poincaré group: spinning particles:

The final example concerns the Poincaré group and the mechanical description of spinning particles.

In the usual terminology, the Lie algebra G of the Poincaré group G is spanned by $\{p_0, p_1, p_2, p_3\}$ (the generators of the translation subgroup) and $\{M_{ab};\ a \neq b = 0, 1, 2, 3\}$ (the generators of the Lorentz subgroup). Since $[G,G] = G$ and $H^2(G;\mathbb{R}) = 0$, the Kostant-Souriau theory can be applied directly.

The polynomials:

$$p^2 = g^{ab}\, p_a\, p_b, \quad w^2 = g_{ab}\, w^a\, w^b \qquad 9.48$$

(where $w^a = \frac{1}{2}\epsilon^{abcd} M_{bc}\, p_d$ and g_{ab} is the Lorentz metric) can be regarded as functions on G^*. It is not hard to see that they are Casimir polynomials (that is, they are preserved under the coadjoint action of G on G^*) and so are constant on the orbits of G.

In fact, the orbits, which represent the possible classical phase spaces of elementary relativistic particles, are completely characterized by the values of p^2 and w^2:

(i) $p^2 > 0$, $w^2 \neq 0$ (massive spinning particles). Putting $m = p^2 \cdot \mathrm{sign}\,(p_0)$ (mass parameter) and $s = -w^2/p^2$ (spin parameter), one has a unique orbit in G^* for every $m \neq 0$ and $s > 0$: each is diffeomorphic with $\mathbb{R}^6 \times S^2$ and is quantizable for integral or half integral values of $2\pi s$. There are two group invariant polarizations (which are complex conjugates of each other). One gives the usual Wigner (m,s)-representations. The other gives nothing.

(ii) $p^2 > 0$, $w^2 = 0$ (massive scalar particles). The orbits are all diffeomorphic with $\mathbb{R}^6$ (and therefore quantizable).

(iii) $p^2 = 0$ (zero rest mass particles). Here the orbits are diffeomorphic with $\mathbb{R}^4 \times S^2$. These are labelled by the energy $\eta = \text{sign}\,(p_o)$ and by the spin s and the helicity χ which are defined by:

$$w_a = \chi s\, p_a \; ; \quad s > 0, \quad \chi = \pm 1 \qquad 9.49$$

(w_a and p_a are necessarily parallel when $p^2 = 0$). Again, the orbits are quantizable when $2\pi s$ is integral or half integral.

The interpretation of these orbits as classical phase spaces, the application of the quantization procedure and the construction of the corresponding irreducible representations of the Poincaré group are described in detail by Souriau, Renouard and Carey[40]. As an example, I shall describe here a slightly different procedure for the zero rest mass (positive energy) case using two component spinors[41] and borrowing some calculations from Penrose's twistor theory[42].

Let O denote a fixed origin in Minkowski space X. In Lorentz coordinates, a (positive energy) zero rest mass particle is characterized by its position vector x^a, its momentum (a future pointing null covector p_a) and its angular momentum about O (a skew tensor M^{ab}). The Pauli-Lubanski vector of the particle

$$w_a = \frac{1}{2} \epsilon_{abcd} p^b M^{cd} \tag{9.50}$$

is necessarily parallel to p_a, whence M^{ab} has the algebraic structure:

$$M^{ab} = \chi s \epsilon^{abcd} p_c n_d + x^a p^b - x^b p^a \tag{9.51}$$

for some null vector n_a, normalized so that:

$$n_a p^a = 1 \quad . \tag{9.52}$$

(Equation 9.43 can be thought of as an alternative definition of the spin s (> 0) and the helicity χ (= ± 1)).

In terms of the variables x^a, p_a and n_a, the symplectic structure is given by the 2-form:

$$\sigma = \chi s \epsilon^{abcd} p_a n_b dp_c \wedge dn_d + dp_a \wedge dx^a \tag{9.53}$$

Though σ is closed, it is degenerate. The phase (M,σ) of the particle is constructed by factoring out the integral manifolds of the kernel. According to Souriau, this amounts to identifying (x^a, p_a, n_a) with $(\tilde{x}^a, \tilde{p}_a, \tilde{n}_a)$ whenever:

$$\begin{aligned} x^a &= \tilde{x}^a + \chi s z^a \\ p_a &= \tilde{p}_a \\ n^a &= \tilde{n}^a + \frac{1}{2}\epsilon^{abcd} \tilde{p}_b \tilde{n}_c z_d + \frac{z^b z_b}{2} . p^a \end{aligned} \tag{9.54}$$

for some z^a such that $p_a z^a = 0$. (Thus a zero rest mass particle cannot be localized: it occupies an entire null hyperplane). Eqn. 9.45 is the only possible choice for σ : it is dictated by the requirement that the phase space should be a homogeneous space for G or, in other words, that (M,σ) should be mapped onto an orbit in G^* as in eqn. 8.11.

In this form, the identification (eqn. 9.46) is not very transparent. It can be rewritten in a simpler form by introducing the spinors corresponding to x^a, p_a and n_a:

$$x^a \longleftrightarrow x^{AA'}$$

$$p_a \longleftrightarrow \bar{\pi}_A \pi_{A'} \qquad 9.55$$

$$n_a \longleftrightarrow \varepsilon_A \bar{\varepsilon}_{A'}$$

where π_A and χ^A are normalized so that

$$\bar{\pi}_A \varepsilon^A = 1 \qquad 9.56$$

There is an arbitrary overall phase factor in the choice of $\pi_{A'}$ and χ^A. This will be factored out later.

If ω^A is defined by:

$$\chi s \varepsilon^A = \omega^A - i x^{AB'} \pi_{B'} \qquad 9.57$$

then σ becomes (after some computation):

$$\sigma = i\,(d\omega^{A} \wedge d\bar{\pi}_{A} + d\pi_{A'} \wedge d\bar{\omega}^{A'}) \qquad 9.58$$

and the identification (eqn. 9.46):

$$x^{AA'} = \tilde{x}^{AA'} + s\chi\, z^{AA'}$$

$$\pi_{A'} = e^{i\theta}\, \tilde{\pi}_{A'} \qquad 9.59$$

$$\omega^{A} = e^{i\theta}\, \tilde{\omega}^{A'}$$

for some real θ and $z^{AA'}$, such that $z^{AA'}\, \bar{\pi}_{A}\, \pi_{A'} = 0$. Thus the particle can be described by the pair $(\omega^{A}, \pi_{A'})$, with the position $x^{AA'}$ given by eqns. 9.48 and 9.49. However, ω^{A} and $\pi_{A'}$ are not arbitrary: again from eqns. 9.48 and 9.49, they must satisfy:

$$\omega^{A}\, \bar{\pi}_{A} + \omega^{A'}\, \pi_{A'} = 2\chi s. \qquad 9.60$$

It follows that the phase space (M,σ) can be constructed in this way: start with the complex four dimensional vector space[43] T of arbitrary pairs $(\omega^{A}, \pi_{A'})$ and the symplectic form[44]:

$$\sigma = i\,(d\omega^{A} \wedge d\bar{\pi}_{A} + d\pi_{A'} \wedge d\bar{\omega}^{A'}) \qquad 9.61$$

Then take the surface $E_{\chi,s}$ given by:

$$h = 2\chi s \qquad 9.62$$

where $h: T \to \mathbb{R} : (\omega, \pi_{A'}) \to \omega^A \bar{\pi}_A + \pi_{A'} \bar{\omega}^{A'}$ and factor out the orbits of ξ_h, that is identify the pairs

$$(\omega^A, \pi_{A'}) \quad \text{and} \quad (\tilde{\omega}^A, \tilde{\pi}_{A'})$$

whenever

$$\omega^A = e^{i\theta}\, \tilde{\omega}^A, \quad \pi_{A'} = e^{i\theta}\, \pi_{A'} \quad ; \quad \theta \in \mathbb{R} \qquad 9.63$$

The factor space is M; σ projects onto the symplectic form on M (also denoted σ).

The quantization of M is now almost trivial. First note that the Poincaré group is generated by functions of the form:

$$(\omega^A, \pi_{A'}) \longmapsto i\,(\lambda^A{}_B\, \omega^B \bar{\pi}_A - \bar{\lambda}^{A'}{}_{B'}\, \pi_A \bar{\omega}^{B'}) \qquad \text{(Lorentz group)}$$

$$(\omega^A, \pi_{A'}) \longmapsto k^{AA'}\, \bar{\pi}_A\, \pi_{A'} \qquad \text{(translation group)}$$

These are at most linear in ω^A and $\bar{\omega}^{A'}$, so the real polarization of T spanned at each point by the vectors $\{\frac{\partial}{\partial \omega^A}, \frac{\partial}{\partial \bar{\omega}^{A'}}\}$ is group invariant. It is also Lie propagated by ξ_h and so projects to a polarization of M.

Next, the prequantization line bundle L over M is constructed by taking the trivial line bundle $E_{\chi,s} \times \mathbb{C}$ and the connection form:

$$\alpha = \theta + \frac{1}{2\pi i} \cdot \frac{dz}{z} \qquad 9.64$$

where

$$\theta = i/_2\,(\omega^A\, d\bar{\pi}_A + \pi_{A'}\, d\bar{\omega}^{A'} - \bar{\omega}^{A'}\, d\pi_{A'} - \bar{\pi}_A\, d\omega^A) \qquad 9.65$$

and factoring out the integral curves of the vector field (on $E_{\chi,s} \times \mathbb{C}$):

$$\zeta = \xi_h - 2\pi i\,(\xi_h \lrcorner\, \theta)\left(z\frac{\partial}{\partial z} - \bar{z}\frac{\partial}{\partial \bar{z}}\right) \qquad 9.66$$

The factor space L is a well defined line bundle over M provided only that the orbits of ζ are closed. One then has:

$$\begin{array}{ccc} E_{\chi,s} \times \mathbb{C} & \longrightarrow & L = (E_{\chi,s} \times \mathbb{C})/\zeta \\ \downarrow & & \downarrow \\ E_{\chi,s} & \longrightarrow & M = E_{\chi,s}/\xi_h \end{array}$$

The condition that the orbits of ζ should be closed is that the integral of θ around each orbit in $E_{\chi,s} \times \mathbb{C}$ should be an integer, that is that $2\pi s$ should be integral or half integral.

When this condition is satisfied, the 1-form α projects into a connection form on L (since $\zeta \lrcorner\, \alpha = 0$ and $\mathcal{L}_\zeta\, \alpha = 0$) with curvature σ (since $\sigma = d\theta$). Thus if $2\pi s$ is integral or half integral, M is quantizable: conversely if M is quantizable then the integral of θ around each orbit of ξ_h in $E_{\chi,s}$ is an integer (proof[45]: the integral of θ around an orbit in $E_{\chi,s}$ is equal to the integral of σ over any 2-surface[46] spanning the orbit: under the projection $E_{\chi,s} \to M$ this

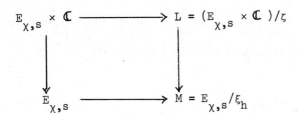

2-surface integral becomes a contour integral in M, since the boundary is mapped to a point; by the quantization condition, the contour integral is an integer) and so $2\pi s$ is integral or half integral.

Assume that $2\pi s$ is integral or half integral.

Any function $\phi: E_{\chi,s} \to \mathbb{C}$ defines a section of $E_{\chi,s} \times \mathbb{C}$. If ϕ satisfies

$$\nabla_{\xi_h} \phi = \xi_h \phi - 2\pi i h \phi = 0 \tag{9.67}$$

then this section will be parallel along ξ_h and will project into a section of L. If additionally:

$$\phi(\omega^A, \bar{\omega}^{A'}, \pi_{A'}, \bar{\pi}_A) = f(\pi_{A'}, \bar{\pi}_A)\, e^{-\pi(\omega^A \bar{\pi}_A - \bar{\omega}^{A'} \pi_{A'})} \tag{9.68}$$

for some function f (independent of ω^A and $\bar{\omega}^{A'}$) then ϕ will project into a polarized section of L. Clearly any polarized section of L can be represented in this way.

Further, the 3-form on T:

$$\nu = \frac{dp_1 \wedge dp_2 \wedge dp_3}{p_o} \tag{9.69}$$

is Poincaré invariant, orthogonal to the polarization and projects to a 3-form on M (ν is, in fact, the standard invariant volume element on the light cone). Thus the polarized wave functions on M can be represented by objects (on $E_{\chi,s}$) of the form:

$$\psi = f \cdot e^{-\pi(\omega^{A}\bar{\pi}_{A} - \bar{\omega}^{A'}\pi_{A'})} \cdot \nu^{\frac{1}{2}} \qquad 9.70$$

where $f : E_{\chi,s} \to \mathbb{C}$ satisfies:

$$\text{(i)} \qquad \frac{\partial f}{\partial \omega^{A}} = \frac{\partial f}{\partial \bar{\omega}^{A'}} = 0 \qquad 9.71$$

$$\text{(ii)} \qquad \xi_{h}\, f = -\, 2\pi i \,.\, 2\chi s \,.\, f \,. \qquad 9.72$$

Eqn. 9.64 can be rewritten as the homogeneity condition:

$$\bar{\pi}_{A}\frac{\partial f}{\partial \bar{\pi}_{A}} - \pi_{A'}\frac{\partial f}{\partial \pi_{A'}} = 4\pi\chi s \,.\, f \qquad 9.73$$

Finally, when $\chi > 0$ the correspondence with the conventional view of a quantized zero rest mass particle can be recovered by introducing the spinor field (with $4\pi s$ indices) on X:

$$\Phi_{A'B'\ldots}(x) = \int_{\pi_{A'},\bar{\pi}_{A}} \pi_{A'}\,\pi_{B'}\,\ldots\, f \; e^{-\pi(\omega^{A}\bar{\pi}_{A} - \bar{\omega}^{A'}\pi_{A'})} \; \nu \qquad 9.74$$

(here ω^{A} and $x^{AA'}$ are related (as before) by eqns 9.48 and 9.49). The homogeneity condition (eqn. 9.65) ensures that $\Phi_{A'B'C'}\ldots$ is well defined.

The exponential factor is simply:

$$e^{-2\pi i\, x^{AA'}\bar{\pi}_{A}\pi_{A'}}$$

so that it follows from eqn. 9.66 that $\Phi_{A'B'C'}\ldots$ obeys the spin s zero-rest-mass field equation:

$$\nabla^{AA'} \Phi_{A'B'\cdots} = 0 \qquad 9.75$$

(where $\nabla_{AA'}$ is the (flat) spinor connection on X). When $\chi < 0$ one must replace $\pi_{A'}$ by $\bar{\pi}_A$ in eqn. 9.66.

A: Čech Cohomology: Chern Classes and Weil's Theorem.

Much of quantization theory is concerned with global properties of symplectic manifolds. In particular, in the prequantization of a symplectic manifold (M,ω), the line bundle needed for the construction of the quantum Hilbert space always exists locally (as the example in §5 shows): the difficulty arises in trying to patch together the small pieces to form a line bundle over the whole of M.

Čech's cohomology theory provides a natural framework within which to discuss the limitations imposed on such patching constructions by the global topology of the underlying manifold. Briefly, it goes like this:

Let M be a smooth manifold and let $\mathcal{U} = \{U_i \mid i \in \Lambda\}$ be a fixed open contractible cover of M (indexed by some set Λ): that is each of the open sets U_i, $U_i \cap U_j$, $U_i \cap U_j \cap U_k$ .... is either empty or can be smoothly contracted to a point (for example, the U_i's might be the normal neighbourhoods of some Riemannian metric).

A <u>k-simplex</u> (in the sense of Čech) is any k + 1-tuple $(i_o, i_1, \ldots, i_k) \in \Lambda^{k+1}$ such that $U_{i_o} \cap U_{i_1} \cap \ldots \cap U_{i_k} \neq \emptyset$.

Let G be an abelian Lie group. For the moment, it will be assumed that G is $\mathbb{Z}$, $\mathbb{R}$, $\mathbb{C}$ or $\mathbb{C}^*$ (that is, the integers or the real or the complex numbers under addition, or the complex numbers under multiplication).

A <u>k-cochain</u> (relative to $\mathcal{U}$) is any totally skew map

$$g: (i_o, i_1, \ldots, i_k) \longmapsto g(i_o, i_1, \ldots, i_k) \in G$$

from the set of k-simplices into G. The set of all k-cochains is denoted $C^k(\mathcal{U},G)$.

The cohomology theory arises from the existence of two natural algebraic structures on the sets of cochains. First, each $C^k(\mathcal{U},G)$ is an abelian group, with the group operation defined by:

$$(g_1 \dot{+} g_2)(i_o, i_1, \ldots, i_k) = g_1(i_o, i_1, \ldots, i_k)$$

$$\dot{+}\, g_2(i_o, i_1, \ldots, i_k); \quad g_1, g_2 \in C^k(\mathcal{U},G) \qquad \text{A.1}$$

(the symbol + or . is used according the context). Second, there is a sequence of group homomorphisms $\delta : C^k(\mathcal{U},G) \to C^{k+1}(\mathcal{U},G)$ defined by:

$$\delta g(i_o,i_1,\ldots,i_{k+1}) = \sum_{j=o}^{k+1} (-1)^i\, g(i_o,i_1,\ldots,\hat{i}_j, \ldots, i_{k+1}); \; G = \mathbb{Z}, \mathbb{R} \text{ or } \mathbb{C} \qquad \text{A.2}$$

$$= \prod_{j=o}^{k+1} (g(i_o,i_1,\ldots,\hat{i}_j, \ldots, i_{k+1})^{(-1)^j} \quad ; \quad G = \mathbb{C}^* \qquad \text{A.3}$$

(^ means "omit"); δ is called the <u>coboundary operator.</u> Note that $\delta^2 : C^k(\mathcal{U},G) \to C^{k+2}(\mathcal{U},G)$ is trivial.

A cochain $g \in C^k(\mathcal{U},G)$ such that $\delta g = 0$ is called a <u>k-cocycle</u>. If, additionally, $g = \delta h$ for some $h \in C^{k-1}(\mathcal{U},G)$ then g is called a <u>k-coboundary</u>. The set of k-cocycles is denoted $Z^k(\mathcal{U},G)$. The k-coboundaries form a subgroup of $Z^k(\mathcal{U},G)$ and the quotient:

$$H^k(\mathcal{U},G) = Z^k(\mathcal{U},G)/\delta(C^{k-1}(\mathcal{U},G)) \qquad \text{A.4}$$

is called the kth cohomology group (relative to $\mathcal{U}$): each element of $H^k(\mathcal{U},G)$ is an equivalence class of cocycles any two of which differ by a coboundary.

It is a standard theorem[47] that $H^k(\mathcal{U},G)$ is independent of $\mathcal{U}$: a different choice of contractible covering will lead to isomorphic cohomology groups. One way of proving this is to construct an isomorphism between $H^k(\mathcal{U},G)$ and a cohomology group constructed from a suitable class of differentiable forms on M (implicitly, a special case of this isomorphism ($k = 2$, $G = \mathbb{R}$) is used in the proof of **Weil's** theorem below). For this reason, one often writes $H^k(M,G)$ for $H^k(\mathcal{U},G)$.

More generally, one can allow $g(i_o, i_1, \ldots i_k)$ to be a smooth function $U_{i_o} \cap U_{i_1} \cap \ldots \cap U_{i_k} \to G$ rather than a fixed element of G. The definitions go through in the same way (eqns. A.1 - A.3 must now be understood as holding at each point). One denotes the resulting cohomology groups: $H^k(\mathcal{U},\underline{G})$. (These cohomology groups are best dealt with using sheaf theory, where the information that the functions must be smooth is encoded in the topological structure of a sheaf over M; this is a technicality which will not be pursued here).

Again, one can further extend these ideas by including non abelian groups. However, though the terminology is useful (for instance, in discussing fibre bundles), the theory is less complete[48]:

for example the group structure of $C^k(\mathcal{U},G)$ does not, in general carry over to $H^k(\mathcal{U},G)$.

<u>Example</u>: Choose a triangulation of M with vertices $\{x_i \mid i \in \Lambda\}$ and take $\mathcal{U}$ to be the set of star neighbourhoods of the vertices, that is $\mathcal{U} = \{U_i\}$ where:

$U_i = \{x \in M \mid x$ lies in the interior of a simplex in the triangulation with vertex $x_i\}$.

Then the incidence relations in the triangulation are reflected in the intersection relations in $\mathcal{U}$, so that:

$x_i, x_j, \ldots\ x_k$ are vertices of a simplex in the triangulation

$$\Leftrightarrow U_i \cap U_j \quad \ldots \quad \cap U_k \neq \emptyset .$$

Thus the simplices making up the triangulation are in one to one correspondence with the Čech simplices of the covering $\mathcal{U}$ (this explains Čech's terminology).

Now let ψ be a complex k-form; ψ defines a k-cochain (relative to $\mathcal{U}$) according to:

$$\psi(i_o, i_1, \ldots i_k) = \int_{x_{i_o} x_{i_1} \ldots x_{i_k}} \psi \qquad \text{A.5}$$

where $x_{i_o}\ x_{i_1}\ \ldots\ x_{i_k}$ is the simplex in the triangulation with vertices

$x_{i_o}, x_{i_1}, \ldots$ and x_{i_k}. Using Stoke's theorem:

$$(d\psi)(i_o, i_1, \ldots, i_{k+1}) = \delta\psi(i_o, i_1, \ldots, i_{k+1}) \qquad \text{A.6}$$

This association of k-forms with k-cochains sets up an isomorphism between the kth (complex) de Rham cohomology group (the set of closed k-forms modulo exact k-forms under addition) and $H^k(\mathcal{U}, \mathbb{C})$. As was remarked above, this isomorphism also exists (though it is harder to exhibit) for more general contractible covers. $\square$.

Now suppose that $\pi : L \to M$ is a line bundle and that $\{U_i, s_i\}$ is a local system for L. By passing, if necessary, to a finer cover, one may suppose that $\mathcal{U} = \{U_i\}$ is contractible. The transition functions:

$$c_{ij} : U_i \cap U_j \to \mathbb{C}^*; \quad U_i \cap U_j \neq \emptyset$$

will then define a 1-cochain $c \in C^1(\mathcal{U}, \mathbb{C}^*)$; c will, in fact, be a cocycle since

$$c_{ij}\, c_{jk}\, c_{ki} = 1 \quad \text{on } U_i \cap U_j \cap U_k \qquad \text{A.7}$$

Thus L determines an equivalence class $[c] \in H^1(\mathcal{U}, \mathbb{C}^*)$. Furthermore, if $\pi_1 : L_1 \to M$ and $\pi_2 : L_2 \to M$ are two equivalent line bundles then the corresponding cohomology classes are equal: by definition, there exists a diffeomorphism $\tau : L_1 \longrightarrow L_2$ commuting with projection

$$\begin{array}{ccc} L_1 & \xrightarrow{\ \tau\ } & L_2 \\ & {\scriptstyle \pi_2} \searrow \quad \swarrow {\scriptstyle \pi_1} & \\ & M & \end{array}$$

and which restricts to a linear isomorphism on each fibre of L_1. By choosing a sufficiently fine contractible covering $\mathcal{U} = \{U_i\}$, one can construct local systems $\{U_i, \underset{(1)}{s}_i\}$ and $\{U_i, \underset{(2)}{s}_i\}$ for L_1 and L_2. If $\{g_i : U_i \to \mathbb{C}^*\}$ is the set of functions defined by:

$$\underset{(1)}{s}_i = g_i \underset{(2)}{s}_i \tag{A.8}$$

then the two sets of transition functions will be related by:

$$\underset{(1)}{c}_{ij} = g_i \underset{(2)}{c}_{ij} g_j^{-1} \tag{A.9}$$

The g_i's define a 0-cochain $g \in C^0(\mathcal{U}, \underline{\underline{\mathbb{C}}}^*)$ and equation A.9 can be rewritten:

$$\underset{(1)}{c} = \delta g \underset{(2)}{c}$$

Thus L_1 and L_2 define the same equivalence class in $H^1(\mathcal{U}, \underline{\underline{\mathbb{C}}}^*)$.

Conversely, given an equivalence class $[c] \in H^1(\mathcal{U}, \underline{\underline{\mathbb{C}}}^*)$ one can construct a line bundle L with transition functions in $[c]$: choose a set of maps:

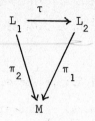

$$c_{ij} : U_i \cap U_j \to \mathbb{C}^* \; ; \quad (i,j) \text{ is a Cech simplex}$$

in [c] (so that the cocycle condition A.7 is satisfied) and take L to be the <u>disjoint</u> union $\bigcup_i U_i \times \mathbb{C}$ factored by the equivalence relation:

$$(x_1,z_1) \sim (x_2,z_2) \; ; \quad x_1 \in U_{i_1} , \; x_2 \in U_{i_2} \; ; \quad z_1,z_2 \in \mathbb{C}$$

whenever:

$$x_1 = x_2 = x \text{ (in M)} \quad \text{and} \quad z_1 = c_{i_1 i_2}(x) \, z_2$$

With the obvious definition of the projection, L is a line bundle over M. One can construct a local system $\{U_i,s_i\}$ for L by defining:

$$s_i : U_i \to L : x \longmapsto [(x,1)] \; ; \quad (x,1) \in U_i \times \mathbb{C}$$

where $[(x,1)] \in L$ is the equivalence class of (x,1) under $\sim$. The transition functions of this local system are simply the c_{ij}'s. Clearly, a different choice of c in [c] will lead to an equivalent line bundle.

Thus, for a given contractible cover $\{U_i\} = \mathcal{U}$, the set $\mathcal{L}$ of equivalence classes of line bundles over M for which it is possible to construct local systems of the form $\{U_i,s_i\}$ is in one to one correspondence with the set of cohomology classes $H^1(\mathcal{U}, \underline{\mathbb{C}}^*)$. (Additionally, this correspondence becomes a group isomorphism when the tensor product[49] is used to define a group structure on $\mathcal{L}$).

The correspondence assumes a more useful, and more readily interpretable, form when $H^1(\mathcal{U}, \mathbb{C}^*)$ is replaced by its natural isomorph $H^2(\mathcal{U}, \mathbb{Z})$ (which is equal to $H^2(M, \mathbb{Z})$ since smooth integer valued functions are necessarily constant). The isomorphism ε: $H^1(\mathcal{U}, \mathbb{C}^*) \to H^2(\mathcal{U}, \mathbb{Z})$ is a consequence of the exact sequence of group homomorphisms:

$$0 \to \mathbb{Z} \to \mathbb{C} \to \mathbb{C}^* \to 0$$

where the map $\mathbb{Z} \to \mathbb{C}$ is inclusion and the map $\mathbb{C} \to \mathbb{C}^*$ is given by: $z \to e^{2\pi i z}$. (Exactness just means that the kernel of each map in the sequence is precisely the image of the preceding map). This induces the exact sequence:

$$\ldots \to H^1(\mathcal{U}, \mathbb{C}) \to H^1(\mathcal{U}, \mathbb{C}^*) \underset{\varepsilon}{\to} H^2(\mathcal{U}, \mathbb{Z}) \to H^2(\mathcal{U}, \mathbb{C}) \to \ldots \to ,$$

implying $H^1(\mathcal{U}, \mathbb{C}^*) \cong H^2(\mathcal{U}, \mathbb{Z})$ since $H^1(\mathcal{U}, \mathbb{C}) = 0$ and $H^2(\mathcal{U}, \mathbb{C}) = 0$ ($\mathbb{C}$ is contractible). If L is a line bundle over M and $\beta(L)$ is the corresponding element of $H^1(\mathcal{U}, \mathbb{C}^*)$ (for some suitable $\mathcal{U}$), then the equivalence class $c(L) = \varepsilon\beta(L) \in H^2(\mathcal{U}, \mathbb{Z}) = H^2(M, \mathbb{Z})$ is called the <u>Chern characteristic class</u> of L.

In less abstract terms, the construction of the Chern class goes as follows: choose a local system $\{U_i, s_i\}$ for L (as above) with transition functions $c_{ij} : U_i \cap U_j \to \mathbb{C}^*$. The Chern class of L is then the equivalence class in $H^2(M, \mathbb{Z})$ of the cocycle:

$$(i,j,k) \longmapsto \frac{1}{2\pi i} \; (\ln c_{ij} + \ln c_{jk} + \ln c_{ki})$$

By the cocycle condition on the c_{ij}'s, the right hand side is a smooth integer valued function on $U_i \cap U_j \cap U_k$, and hence must be constant. There is an ambiguity in taking the logarithms, but the equivalence class in $H^2(M, \mathbb{Z})$ is independent of the branch chosen.

An equivalent, and possibly more intuitive, definition of the Chern class emerges from obstruction theory. Clearly, a line bundle $\pi : L \to M$ is trivial (equivalent to a product) if, and only if, it is possible to find a nowhere vanishing section $s: M \to L$ (since if such a section exists then $M \times \mathbb{C} \to L : (x,z) \longmapsto z.s(x)$ is an equivalence of line bundles). Loosely speaking, one can regard the Chern class as a "measure" of how difficult it is to construct a non-vanishing section. The idea (which works for any bundle) is to choose a triangulation of M (fine enough for the portion of L above each simplex to be trivial) and to try and construct a non-vanishing section $s: M \to L$ by extending s successively from the 0-skeleton (set of 0-simplicies) to the 1-skeleton (set of 1-simplices) and so on until some obstruction is met.

Thus, one first assigns an arbitrary non zero value $s(x_i) \in \pi^{-1}(x_i)$ to each vertex x_i in the triangulation. Extending s to the 1-skeleton is easy: if $x_i x_j$ is a 1-simplex in the triangulation then the portion of L above $x_i x_j$ is diffeomorphic with $x_i x_j \times \mathbb{C}$. To define $s(x)$ for each $x \in x_i x_j$, one simply chooses any curve from $s(x_i)$ to $s(x_j)$ in $x_i x_j \times \mathbb{C}$ which avoids the line $x_i x_j \times \{0\}$. The trouble arises in trying to extend s over the 2-skeleton: the portion of L above a typical 2-simplex $\Delta = x_i x_j x_k$ can again be represented as a product $\Delta \times \mathbb{C}$. The section s has already been defined on the boundary $\partial\Delta$, and this boundary is homeomorphic to a circle S^1. If the image of $s(\partial\Delta)$ under the projection $\pi_2 : \Delta \times \mathbb{C} \to \mathbb{C}$ winds around the origin, then it will be impossible to extend across Δ without allowing s to vanish.

The map which assigns to each 2-simplex Δ (or rather, to the corresponding Čech simplex) the winding number of $\pi_2 \circ s(\partial\Delta)$ around the origin in $\mathbb{C}$ is a cocycle: unless this number can be made to vanish for every simplex in the 2-skeleton, it will be impossible to construct a non-vanishing section and L will be non-trivial.

It turns out that the equivalence class c(L) of this cocycle in $H^2(M, \mathbb{Z})$ is independent of the precise values assigned to s on the 0- and 1-skeletons: it is not hard to see that c(L) is precisely the Chern class of L.

Turning now to Weil's theorem[50] suppose that L is a Hermitian line bundle with connection ∇. As in the example, choose a triangulation of M with vertices $\{x_i\}$ and put U_i equal to the star neighbourhood of x_i. It can be safely assumed that the triangulation is fine enough for there to exist a local system of the form $\{U_i, s_i\}$: in each U_i, the covariant derivative is defined by a 1-form α_i as in eqn. 5.12.

On each non-empty intersection $U_i \cap U_j$, $d(\alpha_i - \alpha_j) = 0$, so that:

$$\alpha_i - \alpha_j = df_{ij}$$

for some function $f_{ij} : U_i \cap U_j \longrightarrow \mathbb{C}$ (since $U_i \cap U_j$ is contractible). By eqn. 5.14, one can put:

$$f_{ij} = \frac{1}{2\pi i} (\ln(c_{ij})$$

for some branch of the logarithm function. Since $d(f_{ij} + f_{jk} + f_{ki}) = 0$, the function:

$$a(i,j,k) = f_{ij} + f_{jk} + f_{ki} : U_i \cap U_j \cap U_k \rightarrow \mathbb{C}$$

must be constant for each Čech 2-simplex (i,j,k): it was pointed out above that this constant is always an integer.

Now the cohomology class of ω in $H^2(M,\mathbb{C})$ is given by the integral:

$$\omega(i,j,k) = \int_{\Delta_{ijk}} \omega$$

over each 2-simplex $\Delta_{ijk} = x_i x_j x_k$ in the triangulation. After a short calculation, using Stoke's theorem:

$$\omega(i,j,k) = \frac{1}{6} (2 \oint_{\partial\Delta_{ijk}} \alpha_i + \oint_{\partial\Delta_{ijk}} \alpha_j + \oint_{\partial\Delta_{ijk}} \alpha_k)$$

$$= \frac{1}{6} \; [2(f_{ij} + f_{jk} + f_{ki}) \, (x_i) + 2(f_{ij} + f_{jk} + f_{ki}) \, (x_j) +$$

$$+ \quad 2(f_{ij} + f_{jk} + f_{ki})\,(x_k)]$$

$$- \frac{1}{2}\;[(f_{ij}(x_i) + f_{ij}(x_j)) + (f_{jk}(x_j) + f_{jk}(x_k)) + (f_{ki}(x_k) + f_{ki}(x_i))]$$

$$- \frac{1}{2}\;\Big|\int_{x_i x_j} (\alpha_i + \alpha_j) + \int_{x_j x_k} (\alpha_j + \alpha_k) + \int_{x_k x_i} (\alpha_k + \alpha_i)\Big|$$

The first square bracket is the cocycle $a(i,j,k)$: by definition, this lies in the Chern class of L in $H^2(M,\mathbb{Z})$. The second two square brackets are simply coboundaries. Thus the cohomology class of ω is equal to the Chern class of L. In particular ω must be integral, that is its cohomology class lies in the image of the natural homomorphism

$$\iota \; : H^2(M,\mathbb{Z}) \to H^2(M,\mathbb{C}),$$

induced by the injection $\mathbb{Z} \hookrightarrow \mathbb{C}$.

Conversely, suppose that ω is a real closed 2-form in $\iota(H^2(M,\mathbb{Z})) \subset H^2(M,\mathbb{R})$. Triangulating M as above, on each star neighbourhood U_i:

$$\omega = d\,\alpha_i$$

for some 1-form α_i (since U_i is contractible) and on each $U_i \cap U_j \neq \emptyset$:

$$\alpha_i - \alpha_j = df_{ij}$$

for some function $f_{ij} : U_i \cap U_j \to \mathbb{R}$. Again, since $d(f_{ij} + f_{jk} + f_{ik}) = 0$, $a(i,j,k) = f_{ij} + f_{jk} + f_{ki}$ is constant on $U_i \cap U_j \cap U_k$ and the two cocycles:

$$(i,j,k) \longmapsto \int_{x_i x_j x_k} \omega$$

$$(i,j,k) \longmapsto a(i,j,k)$$

define the same equivalence class in $H^2(M,\mathbb{R})$. But, by hypothesis, there is some cocycle b in this equivalence class such that $b(i,j,k)$ is an integer for each 2-simplex (i,j,k); thus:

$$b = a + \delta g$$

for some $g \in H^1(M,\mathbb{R})$. If $\tilde{f}_{ij} = f_{ij} + g(i,j)$ then:

$$c_{ij} = \exp(2\pi i \, \tilde{f}_{ij}) : U_i \cap U_j \to \mathbb{C}^*$$

will be a cocycle in $H^1(M, \underline{\underline{\mathbb{C}}}^*)$ since:

$$c_{ij}c_{jk}c_{ki} = \exp\left[2\pi i \, (\tilde{f}_{ij} + \tilde{f}_{jk} + \tilde{f}_{ki})\right]$$

$$= \exp\left[2\pi i \, (b(i,j,k))\right]$$

$$= 1 .$$

As above, one can construct a line bundle L using the c_{ij}'s as transition functions : if the α_i's are used to define a connection ∇ on L, then $\operatorname{curv}(L,\nabla) = \omega$. Further, since the transition functions all map into the unit circle in $\mathbb{C}^*$, L will have a natural Hermitian metric which can easily be seen to be ∇-invariant.

The only freedom one has in this construction is to replace c_{ij} by:

$$\tilde{c}_{ij} = c_{ij}\, h(i,j)$$

where $h \in C^1(M,T)$ is a cocycle and T is the circle group (the subgroup of $\mathbb{C}^*$ of complex numbers of unit modulus). If M is simply connected, $H^1(M,T) = \{1\}$ and L is unique up to equivalence; otherwise the possible equivalence classes of line bundles will be parametrized by $H^1(M,T)$. The different possibilities correspond to choosing different b's which define the same cohomology class in $H^2(M,\mathbb{R})$ but different cohomology classes in $H^2(M,\mathbb{Z})$.

B: Principal Bundles and the Existence of Metalinear Structures.

The purpose here is to demonstrate the existence conditions for the metalinear frame bundles used in quantization and to explain some of the related terminology.

First, a few definitions. The line bundles encountered in §5 were special cases of a more general concept, that of a bundle over a manifold. Explicitly, $\pi : B \to M$ is a bundle over M if

1) B and M are C^∞ manifolds (B is called the bundle space, and M the base space)
2) π is a smooth map of B onto M (π is called the projection).

The counter image $B_x = \pi^{-1}(x)$ of x M is called the fibre over x. If the fibres are submanifolds and if each is diffeomorphic with some fixed manifold F, then $\pi : B \to M$ is called a fibre bundle with fibre F. A fibre bundle is usually denoted either as a quadruple (B,π,M,F) or, where there is no possible ambiguity, by its bundle space B.

A local trivialization of a fibre bundle (B,π,M,F) is a pair (ϕ,U) where U is an open contractable subset of M and

$$\phi : U \times F \to \pi^{-1}(U) \subset B$$

is a local diffeomorphism which commutes with projection:

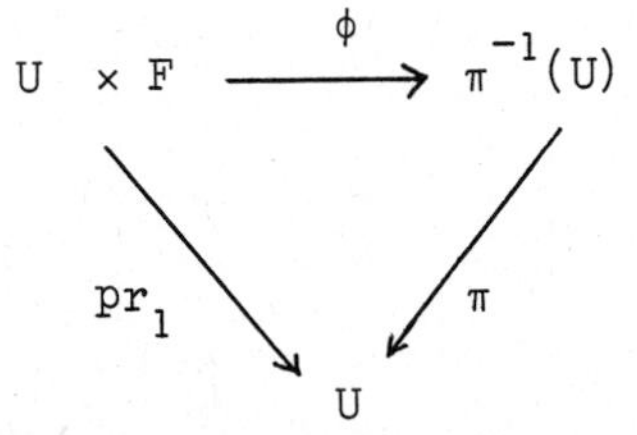

Here $pr_1 : U \times F \to U$ is the projection onto the first factor. If there exists a collection $C = \{(\phi_i, U_i) \mid i \in I\}$ (indexed by some set I) of local trivializations such that $\{U_i \mid i \in I\}$ covers M then (B,π,M,F) is said to be locally trivial; C is called an <u>atlas</u> for the bundle.

If there is given a Lie group G which acts effectively[51] on F as a group of diffeomorphisms then one says that two local trivializations (Φ,U) and (Ψ,V) are G-compatible either if $U \cap V = \emptyset$ or if there exists a smooth map $g: U \cap V \to G$ such that, for each $b \in \pi^{-1}(U \cap V)$:

$$pr_2(\phi^{-1}(b)) = g(\pi(b)).(pr_2(\psi^{-1}(b)))$$

where $pr_2 : (U \cap V) \times F \to F$ is the projection onto the second factor; g is called the <u>transition function</u> between (ϕ,U) and (ψ,V). The condition that the group action is effective ensures that g is unique if it exists.

A G-<u>atlas</u> for a fibre bundle is an atlas C_G any two elements of which are G-compatible; C_G is complete if it contains all local trivializations which are G-compatible with all elements of C_G. Any G-atlas can be extended to a complete G-atlas.

Finally, a <u>fibre bundle with structure group</u> G is a locally trivial fibre bundle (B,π,M,F) together with a fixed complete G-atlas C_G; the elements of C_G are called <u>admissable local trivializations</u>.

A few examples should clarify these ideas:

1) A line bundle $\pi: L \to M$ is a fibre bundle with fibre $\mathbb{C}$ and structure group $\mathbb{C}^*$. Any non vanishing local section $s: U \subset M \to L$ defines a local trivialization.

2) The tangent bundle of an n-dimensional manifold X is a fibre bundle with fibre $\mathbb{R}^n$ and structure group $GL(n,\mathbb{R})$ (The group of non-singular $n \times n$ real matrices). Any set $\{\xi_1 \ldots \xi_n\}$ of vector fields which are linearly independent in some open set in X defines a local trivialization.

3) The complexified tangent bundle $TX^{\mathbb{C}}$ has fibre $\mathbb{C}^n$ and structure group $GL(n,\mathbb{C})$.

4) A polarization P of a symplectic manifold (M,ω) can be thought of as a fibre bundle: its bundle space is

$$\bigcup_{x \in M} \{x\} \times P_x \subset TM^{\mathbb{C}}$$

its fibre is $\mathbb{C}^n$ and its structure group is $GL(n,\mathbb{C})$. More precisely, P is a sub-bundle of $TM^{\mathbb{C}}$.

Two fibre bundles[52] over M, (B_1,π_1,M,F_1) and (B_2,π_2,M,F_2) with the same structure group G are said to be <u>isomorphic</u> or <u>equivalent</u> if there exist diffeomorphisms:

$$\varepsilon : B_1 \to B_2 \,, \quad \sigma : F_1 \to F_2$$

such that:

1) ε commutes with projection:

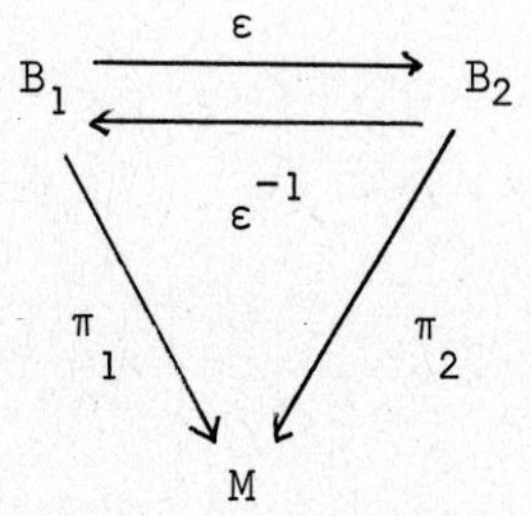

2) (ϕ, U) is an admissable local trivialization of (B_1, π_1, M, F_1) if, and only if, $(\rho \circ \phi \circ \tilde{\sigma}, U)$ is an admissable local trivialization of (B_2, π_2, M, F_2) where:

$$\tilde{\sigma} : U \times F_1 \to U \times F_2 : (x, f_1) \longmapsto (x, \sigma(f_2))$$

In general, the structure group of a fibre bundle[52] does not appear as an explicit part of the geometry: though one can pick out a preferred class of fibre preserving diffeomorphisms of the bundle space which coincide, in any local trivialization, with the action of some element of the structure group on each fibre, there is, in general, no way of associating a given element of the structure group with a particular transformation of the bundle space.

A familiar example is provided by the tangent bundle TX of a manifold X. Each non singular tensor field S of type $\binom{1}{1}$ defines a diffeomorphism of TX which preserves the fibres. The action of S is linear so that in any local trivialization S can be identified on each fibre with an element of $GL(n,\mathbb{R})$. However, there is no canonical way of associating a tensor field S with a given element of $GL(n,\mathbb{R})$, nor, conversely, is there a canonical way of representing a given tensor S as a field of $n \times n$ matrices.

It is this that makes the concept of a principal fibre bundle so useful. A principal bundle is one in which the fibre and the group coincide, with the group acting on itself by left translation. (One also applies this term to a bundle which is isomorphic with a principal bundle). Every fibre bundle can be given an alternative representation in terms of an associated principal bundle, with the structure group appearing explicitly as a transformation group of the associated bundle space.

The construction of the associated bundle depends on the fact that a fibre bundle is completely determined (up to equivalence) by its fibre and by the transition functions:

$$g_{ij} : U_i \cap U_j \to G$$

of some atlas of admissable local trivializations. Just as in the line bundle case, these functions satisfy the cocycle condition:

$$g_{ij}\, g_{jk} = g_{ik} \qquad \text{on} \qquad U_i \cap U_j \cap U_k$$

so that they can be thought of as defining an element $[g]$ of $H^1(M,\underline{G})$: the reconstruction of the bundle from the cohomology class $[g]$ is the same as in the line bundle case.

Given a fibre bundle (B,π,M,F) with structure group G, one obtains the associated principal bundle by choosing an admissable G-atlas, replacing F by G and applying this construction. Conversely, one can reverse the procedure and recover a bundle from its associated principal bundle: one simply has to specify the fibre and the group action.

The structure group G acts on the bundle space B of a principal bundle by right translation: for each $h \in G$, $h: B \to B$ is defined by:

$$h(b) = \phi(\pi(b),gh) \ ; \qquad b \in B$$

where $\phi: U \times G \to \pi^{-1}(U) \ni b$ is an admissable local trivialization and $(\pi(b),g) = \phi^{-1}(b)$. The definition is natural (independent of ϕ) and the action preserves the fibres of B. The corresponding definition of left translation does not make sense unless G is abelian, in which case left and right translation coincide.

A second useful property of principal bundles is that their local admissable trivializations can be described in a very simple way. Explicitly, a <u>local</u> <u>section</u> of a bundle $\pi: B \to M$ is a map $s: U \subset M \to B$ (on some open set $U \subset M$) which commutes with projection:

$$\begin{array}{c} B \\ \pi \downarrow \uparrow s \\ M \end{array}$$

If (B,π,M,G) is a principal bundle and $s: U \subset M \to B$ is a local section then

$$\phi : U \times G \to B: (x,g) \longmapsto g \circ s(x)$$

is a local admissable trivialization; here $g: B \to B$ is the right translation by g. Conversely, a given admissable local trivialization $\phi: U \times G \to B$ can be recovered in this way from the local section:

$$s : U \to B : x \longmapsto \phi(x,e)$$

where $e \in G$ is the identity.

Because of this, it is common to call a local section of a principal bundle a local trivialization. A covering of M by local sections is called a <u>local system</u>.

<u>Example</u>: A vector bundle is a fibre bundle with a vector space as fibre and the appropriate general linear group as structure group. The associated principal bundle of a vector bundle (B,π,M,V^n) (V^n is an n dimensional vector space) is isomorphic with the <u>frame bundle</u> $\tilde{\pi}: \tilde{B} \to M$; each point of $\tilde{B}$ is an $(n+1)$-tuple $(x, \xi_1 \ldots \xi_n)$ where $x \in M$ and $\{\xi_1 \ \ldots \ \xi_n\}$ is a basis for $\pi^{-1}(x)$. A local trivialization ϕ is defined by choosing n local sections $\zeta_1, \ldots \zeta_n : U \to B$ of the vector bundle such that $\zeta_1(x) \ldots \zeta_n(x)$ are linearly independent at each point $x \in U$ and putting:

$$\phi : U \times G \to \tilde{B}: \quad (x, g_{ij}) \longmapsto \zeta_i(x)\ g_{ij}$$

Right translations of $\tilde{B}$ correspond to making the same change of basis at each point of M.

The problem posed in §6 concerned the frame bundle B^P of a polarization P of a symplectic manifold (M,ω): P defines a vector bundle over M with structure group $GL(n,\mathbb{C})$ and B^P is the associated principal bundle.

In the language introduced above, the problem is this: given the principal $GL(n,\mathbb{C})$ bundle

$$\pi : B^P(M) \to M$$

is it possible to find a principal $ML(n,\mathbb{C})$ bundle $\tilde{\pi}: \tilde{B}^P(M) \to M$ and a double covering $\rho : \tilde{B}^P(M) \to B^P(M)$ such that:

1) ρ commutes with projection:

$$\begin{array}{ccc} \tilde{B}^P(M) & \xrightarrow{\ \rho\ } & B^P(M) \\ & {\scriptstyle\tilde{\pi}}\searrow \quad \swarrow{\scriptstyle\pi} & \\ & M & \end{array}$$

2) ρ commutes with right translation:

$$\begin{array}{ccc}
\tilde{B}^P(M) \times ML(n,\mathbb{C}) & \rightarrow & \tilde{B}^P(M) \\
\Big\downarrow \rho \times \sigma & & \Big\downarrow \rho \\
B^P(M) \times GL(n,\mathbb{C}) & \rightarrow & B^P(M)
\end{array}$$

where the horizontal arrows are the group actions and $\sigma\colon ML(n,\mathbb{C}) \to GL(n,\mathbb{C})$ is the group covering map?

This can be restated: given a contractible open cover $\{U_i\}$ of M and the set of transition functions:

$$g_{ij} : U_i \cap U_j \to GL(n,\mathbb{C})$$

of B^P, is it possible to find a set of maps:

$$z_{ij} : U_i \cap U_j \to \mathbb{C}^*$$

such that:

1) $\quad \Delta g_{ij} = (z_{ij})^2 \qquad \text{on} \qquad U_i \cap U_j$

2) $\quad z_{ij}\, z_{jk} = z_{ik} \qquad \text{on} \qquad U_i \cap U_j \cap U_k$.

If this is possible, then the maps:

$$\tilde{g}_{ij} : U_i \cap U_j \to ML(n,\mathbb{C}) : x \longmapsto \begin{bmatrix} g_{ij}(x) & 0 \\ 0 & z_{ij}(x) \end{bmatrix}$$

will satisfy the cocycle condition and they can be used to construct $\tilde{B}^P$.

Proceeding naively, for each $g_{ij} : U_i \cap U_j \to GL(n,\mathbb{C})$, take $f_{ij} : U_i \cap U_j \to \mathbb{C}^*$ to be one of the square roots of Δg_{ij}. There is no difficulty here, since $U_i \cap U_j$ is contractible. Then:

$$(f_{ij}\, f_{jk}\, f_{ki})^2 = 1$$

Thus, if $a(i,j,k) = f_{ij}\, f_{jk}\, f_{ki}$, then:

$$a: (i,j,k) \longmapsto a(i,j,k) \in \mathbb{Z}_2$$

is a cocycle and hence defines an equivalence class $[a] \in H^2(M,\mathbb{Z}_2)$ ($\mathbb{Z}_2$ is the group { + 1, - 1 } under multiplication). This equivalence class is independent of the choice of the f_{ij}'s since, if $\tilde{f}_{ij} : U_i \cap U_j \to \mathbb{C}^*$ is a different choice then:

$$\tilde{a}(i,j,k) = \tilde{f}_{ij}\, \tilde{f}_{jk}\, \tilde{f}_{ki} = a(i,j,k)\; b(i,j)\; b(j,k)\; b(k,i)$$

is equivalent to a, where:

$$b(i,j) = \tilde{f}_{ij} / f_{ij} : U_i \cap U_j \to \mathbb{Z}_2 \quad .$$

By construction, a is a coboundary in $C^2(\mathcal{U}, \mathbb{C}^*)$ ($\mathcal{U}$ is the open cover $\{U_i\}$) but, in general, a will not be a coboundary in $C^2(\mathcal{U}, \mathbb{Z}_2)$. If a is a coboundary in $C^2(\mathcal{U}, \mathbb{Z}_2)$ then, for each Cech simplex (i,j,k):

$$a(i,j,k) = c(i,j)\ c(j,k)\ c(k,i)$$

for some cochain $c \in C^1(\mathcal{U}, \mathbb{Z}_2)$ and the problem is solved by putting:

$$z_{ij} = f_{ij}/c_{ij}$$

Conversely, if z_{ij}'s can be found then, on making the choice $f_{ij} = z_{ij}$, it is clear that [a] vanishes (that is, that [a] is the equivalence class of the cocycle:

$$(i,j,k) \longmapsto 1$$

in $H^2(M, \mathbb{Z}_2)$.

Thus the construction of $\tilde{B}^P(M)$ is possible if and only [a] vanishes in $H^2(M, \mathbb{Z}_2)$; a is called the <u>obstruction cocycle</u>.

If the obstruction does vanish, then the only freedom available in the construction of $\tilde{B}^P(M)$ is in the choice of c. Just as in the proof of Kostant's theorem, the various possible equivalence classes of $\tilde{B}^P(M)$ will be parameterized by the various possible choices for [c] in $H^1(M, \mathbb{Z}_2)$.

<u>Remark</u>: The z_{ij}'s are the transition functions for L^P.

C: Lie Algebra Cohomology and Central Extensions:

This appendix is concerned with a detailed examination of the problem raised in §8: given a Lie group G, with Lie algebra $\mathcal{G}$, a symplectic manifold (M,ω) and a Lie algebra homomorphism

$$\sigma : \mathcal{G} \to A(M)$$

is it possible to find a lifting of σ :

$$\lambda : \mathcal{G} \to C^{\infty}_{\mathbb{R}}(M)$$

which makes

$$\begin{array}{ccc} & \overset{\lambda}{\nearrow} & C^{\infty}_{\mathbb{R}}(M) \\ \mathcal{G} & & \downarrow \\ & \underset{\sigma}{\searrow} & A(M) \end{array}$$

into a commutative diagram of Lie algebra homomorphisms? The answer depends on the structure of $\mathcal{G}$ and, in particular, on its real cohomology.

The cohomology theory of Lie algebras[53] bears a strong formal resemblance to Čech theory and, in fact, the topological structure of a Lie group is closely related to the cohomology of its Lie algebra (see, for example, Chevalley and Eilenberg[53]). However, all that is needed here are some basic definitions: the formal resemblance will be emphasized but not explained.

Let L denote a real Lie algebra. A totally skew multilinear map:

$$g : L^k = L \times L \times \dots \times L \to \mathbb{R}$$

is called a k-cochain and the vector space of all k-cochains is denoted $C^k(L, \mathbb{R})$.

The Lie bracket in L gives rise to a series of linear maps $\delta : C^k(L, \mathbb{R}) \to C^{k+1}(L, \mathbb{R})$ given by:

$$\delta f(X_1 \dots X_{k+1}) = \sum_{i<j} (-1)^{i+j} f([X_i, X_j], X_1 \dots \hat{X}_i \dots \hat{X}_j \dots X_{k+1})$$

(here $f \in C^k(L; \mathbb{R})$ and $X_i \in L$; for $k = 0$ one takes $\delta f = 0$). In the same terminology as was used in Cech theory, δ is called the <u>coboundary operator</u>; as before $\delta^2 = 0$. A cochain f such that $\delta f = 0$ is called a <u>cocycle</u>. If, additionally, $f = \delta g$ for some g then f is called a <u>coboundary</u>. The quotient $H^k(L, \mathbb{R})$ of the space $Z^k(L, \mathbb{R})$ of k-cocycles by the space $\delta(C^{k-1}(L, \mathbb{R}))$ of k-coboundaries is called the kth cohomology group of L. Each element of $H^k(L, \mathbb{R})$ is an equivalence class ("cohomology class") of cocycles, any two of which differ by a coboundary.

The usefulness of all this is that, first, the existence of the lifting $\lambda : G \to C^\infty_{\mathbb{R}}(M)$ can be reduced to the vanishing of a certain equivalence class in $H^2(G, \mathbb{R})$ associated with σ and second that there are a number of standard techniques for computing the cohomology groups of the Lie algebras encountered in physics. For example, for semi-simple Lie algebras (the case dealt with below), $H^2(G,\mathbb{R}) = \{0\}$ and λ always exists.

The construction of the cohomology class associated with σ goes like this: first choose an arbitrary linear map

$$\lambda_o : G \to C^\infty_{\mathbb{R}}(M)$$

such that

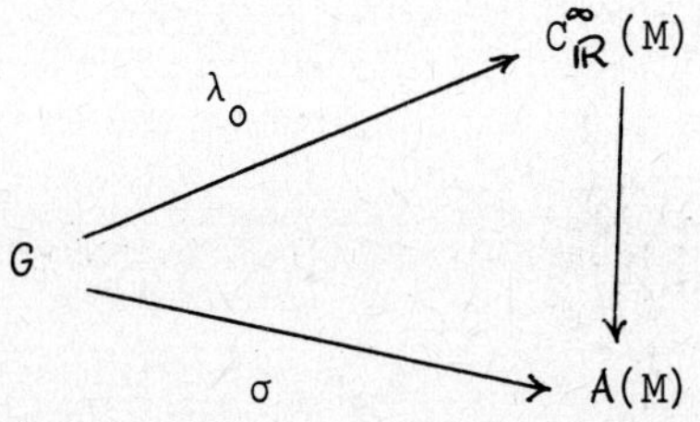

commutes (in general, λ_o will not be a Lie algebra homomorphism). Next put:

$$f(X,Y) = [\lambda_o(X),\lambda_o(Y)] - \lambda_o([X,Y]) \in C^\infty_{\mathbb{R}}(M) \qquad \text{C.1}$$

where $X,Y \in G$ (the first bracket is the Poisson bracket, the second is the Lie bracket in G); crudely, f measures by how much λ_o fails to be a Lie algebra homomorphism. Since σ preserves brackets, $\xi_{f(X,Y)} = 0 \in A(M)$, so that $f(X,Y)$ is constant for each $X,Y \in G$.

Also, from the Jacobi identities in G and $C^{\infty}_{\mathbb{R}}(M)$, for any $X,Y,Z \in G$

$$f([X,Y],Z) + f([Z,X],Y) + f([Y,Z],X) = 0 \qquad \text{C.2}$$

implying that $f: G \times G \to \mathbb{R}$ is a cocycle. It is not hard to see that the cohomology class $[f] \in H^2(G; \mathbb{R})$ is characteristic of σ, since the only freedom in constructing f is to replace λ_o by $\lambda_o + h$ where

$$h : G \to \mathbb{R}$$

is linear. The result is to replace f by $f - \delta h$, leaving $[f]$ unaltered.

If λ exists then, on taking $\lambda = \lambda_o$, one obtains $f = 0$ and the cohomology class vanishes. Conversely, if $[f] = 0$ then $f = \delta g$ for some $g \in C^1(G, \mathbb{R}) = G^*$. Replacing λ_o by $\lambda = \lambda_o + g$ gives:

$$\begin{aligned}[\lambda(X),\lambda(Y)] &= [\lambda_o(X) + g(X),\ \lambda_o(Y) + g(Y)]; \quad X,Y \in G \\ &= [\lambda_o(X),\ \lambda_o(Y)] \\ &= \lambda_o([X,Y]) + g([X,Y]) \\ &= \lambda([X,Y]) \qquad \text{C.3}\end{aligned}$$

implying that λ is a Lie algebra homomorphism. To summarize, λ

exists if and only if $[f] = 0$.

It was remarked in §4 that essentially the same lifting problem arises in quantum mechanics. This can now be clarified. The conventional group theoretic approach to the quantization of a classical system with invariance group G is based on the symmetry axiom, which states that the group G must also act irreducibly **as a** symmetry group on the quantum phase space. This phase space is the set $\hat{H}$ of rays in the quantum Hilbert space H . The action of each group element $g \in G$ on $\hat{H}$ can be represented by any one of a family of unitary transformations of H, any two of which differ by an overall phase factor. The question of whether it is possible to choose a representative element of each family so as to obtain a unitary representation of G on H can be reduced in precisely the same way to a question concerning the cohomology of G. The difference between the conventional approach and the Kostant-Souriau quantization scheme is that the lifting problem is first encountered at the classical rather than at the quantum level (for more details, see Simms[54]).

The usefulness of the lifting criterion is illustrated by its application to semi-simple Lie algebras (of which the Lie algebra of SO(3) is an example): a Lie algebra L is said to be <u>semi-simple</u> if its Killing form is non-degenerate. That is if the symmetric bilinear form $\langle , \rangle$ given by

$$\langle , \rangle : L \times L \to \mathbb{R} : (X,Y) \mapsto \langle X,Y \rangle = \mathrm{tr}(\mathrm{ad}\, X . \mathrm{ad}\, Y)$$

is nonsingular (here ad is the adjoint representation of L on L defined by

$$\text{ad } X : L \to L : Y \to [X,Y].)$$

An equivalent characterization of a semi-simple Lie algebra[55] is that it should contain no abelian ideals other than {0}.

In any semi-simple Lie algebra L, $H^2(L,\mathbb{R}) = 0$. To prove this, first note that if L is semi-simple then[56]

$$\begin{aligned} <[X,Y],Z> &= \text{tr } [(\text{ad } X \text{ ad } Y - \text{ad } Y.\text{ad } X) \text{ ad } Z] \\ &= \text{tr } [\text{ad } X (\text{ad } Y.\ \text{ad } Z - \text{ad } Z\ .\ \text{ad } Y)] \\ &= <X, [Y,Z]> \quad ; \quad X,Y,Z \in L. \end{aligned} \qquad \text{C.5}$$

Next, every linear map g: L → L which satisfies

$$g([X,Y]) = [g(X),Y] + [X,g(Y)] \quad \forall \ X,Y \in L \qquad \text{C.6}$$

must be of the form g = ad Z for some Z ∈ L (in more technical terms: every derivation of L is an inner derivation). This follows from the non degeneracy of <,> since the linear map

$$L \to \mathbb{R} : \quad X \longmapsto \text{tr } (g\ .\ \text{ad } X)$$

is an element of the dual space L^* and so must be of the form

$$X \mapsto \langle Z, X\rangle \quad ; \qquad X \in L$$

for some $Z \in L$. Thus

$$\mathrm{tr}(g \,.\, \mathrm{ad}\, X) = \langle Z, X\rangle = \mathrm{tr}(\mathrm{ad}\, Z \,.\, \mathrm{ad}\, X) \;\forall\; X \in L \qquad \text{C.7}$$

and so:

$$\mathrm{tr}((g - \mathrm{ad}\, Z) \,.\, \mathrm{ad}\, X) = 0 \qquad \forall \quad X \in L. \qquad \text{C.8}$$

Putting $u = g - \mathrm{ad}\, Z$, it must be shown that $u = 0$. Now, from eqn. C.6 and the Jacobi identity:

$$(\mathrm{ad}\, u(X))Y = [(g - \mathrm{ad}\, Z)X, Y] \quad ; \qquad X, Y \in L$$

$$= g([X,Y]) - [X, g(Y)] - [[Z,X],Y]$$

$$= u(\mathrm{ad}\, X(Y)) - \mathrm{ad}\, X(u(Y)) \qquad \text{C.9}$$

implying:

$$\mathrm{ad}\, u(X) = u \,.\, \mathrm{ad}\, X - \mathrm{ad}\, X \,.\, u \qquad \text{C.10}$$

Hence, for any $X, Y \in L$:

$$\langle Y,u(X)\rangle = tr(ad\, Y \,.\, ad\, u(X))$$

$$= tr(ad\, Y(u \,.\, ad\, X - ad\, X \,.\, u))$$

$$= tr(u(ad\, X \,.\, ad\, Y - ad\, Y \,.\, ad\, X))$$

$$= tr(u \,.\, ad\, [X,Y])$$

$$= 0 \qquad C.11$$

implying $u(X) = 0 \quad \forall \quad X \in L$.

Finally, suppose that $f \in C^2(L,\mathbb{R})$ is a cocycle. That is, for any $X,Y\ Z \in L$:

$$f([X,Y],Z) + f([Y,Z],X) + f([Z,X],Y) = 0 \qquad C.12$$

Now, for fixed $X \in L$, the map $Y \longmapsto f(X,Y)$ is an element of L^* and so is necessarily of the form

$$Y \longmapsto \langle g(X),Y\rangle$$

for some $g(X) \in L$. Clearly $X \mapsto g(X)$ is linear. Also, using eqn. C.12, g satisfies

$$g([X,Y]) = [g(X),Y] + [X,g(Y)] \quad \forall \quad X,Y \in L \qquad C.13$$

Thus $g = \text{ad } Z$ for some $Z \in L$ and so:

$$f(X,Y) = \langle g(X),Y\rangle$$

$$= \langle [Z,X],Y\rangle$$

$$= \langle Z,[X,Y]\rangle \qquad \text{C.14}$$

That is $f = \delta h$ where:

$$h : L \to \mathbb{R} : X \longmapsto \langle Z,X\rangle$$

and $H^2(L,\mathbb{R}) = 0$.

When $H^2(G,\mathbb{R})$ does not vanish (as in the case of the Galilei group) one can still achieve the lifting by passing to a central extension of G. A Lie algebra E is called a central extension[57] of a given algebra L (by $\mathbb{R}$) if there exists an exact sequence of Lie algebras:

$$0 \to \mathbb{R} \to E \overset{\pi}{\to} L \to 0$$

In other words, if there exists a homomorphism $\pi: E \to L$ of E onto L with kernel isomorphic to $\mathbb{R}$. The extension is <u>trivial</u> if the sequence <u>splits</u>, that is if it is possible to find an injection $\iota : L \to E$ which commutes with π; diagrammatically:

$$0 \to \mathbb{R} \to E \underset{\iota}{\overset{\pi}{\rightleftarrows}} L \rightleftarrows 0$$

This situation is trivial in the sense that E is then the direct sum (as a vector space) of two subalgebras: the one $(\iota(L))$ isomorphic with L, the other with $\mathbb{R}$ (as a trivial Lie algebra)[58].

There is a natural concept of equivalence between central extensions of a given algebra L: $\pi : E \to L$ and $\tilde{\pi} : \tilde{E} \to L$ are equivalent if there exists a Lie algebra homomorphism $\phi : E \to \tilde{E}$ such that

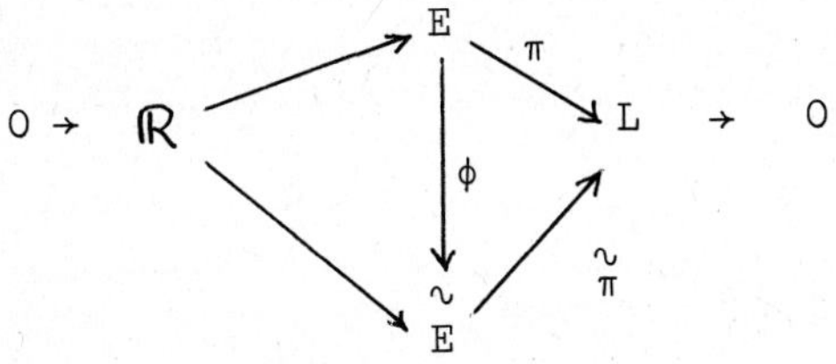

commutes.

The point of this is that the inequivalent central extensions of L are parameterized by $H^2(L,\mathbb{R})$: with each extension $\pi : E \to L$, there is associated an equivalence class $[f] \in H^2(L,\mathbb{R})$ where

$$f(X,Y) = [h(X),h(Y)] - h[X,Y] \qquad ; \qquad X,Y \in L$$

and $h: L \to E$ is any linear map commuting with π ($f(X,Y) \in \mathbb{R}$ since $\pi(f(X,Y)) = 0$: here $\mathbb{R}$ is identified with the kernel of π). Conversely, given an equivalence class $[f] \in H^2(L,\mathbb{R})$, one can construct a central extension of L by taking $E = L \oplus \mathbb{R}$ (as a vector space) with the Lie bracket

$$[X + r, \; Y + s] = [X,Y] + f(X,Y); \quad X,Y \in L; \quad r,s \in \mathbb{R} \qquad \text{C.15}$$

together with the obvious definition of π. This will be trivial if and only if f is a coboundary: in fact if $f = \delta g$ then the splitting is given by:

$$\iota(X) = X + g(X) \in L \oplus \mathbb{R} \quad ; \quad X \in L. \qquad \text{C.16}$$

Returning to the lifting problem, when $[f] \neq 0$, one can still construct λ by passing first to the central extension $E = G \oplus \mathbb{R}$ (with the Lie bracket as in eqn. C.15) of L defined by f. For if $\tilde{\sigma} : E \to A(M)$ is defined by:

$$\tilde{\sigma}(X + r) = \sigma(X); \quad X + r \in E$$

then the cohomology class $[\tilde{f}]$ of $\tilde{\sigma}$ is related to f by:

$$\tilde{f}(X + r, Y + s) = f(X,Y); \quad X + r, \; Y + s \in E$$

But in $H^2(E,\mathbb{R})$, $[\tilde{f}] = 0$: in fact $\tilde{f} = \delta\tilde{g}$ where

$$\tilde{g} : E \to \mathbb{R} : X + r \longrightarrow r.$$

Notes

1. C.W. Misner, K.S. Thorne and J.A. Wheeler: Gravitation (Freeman, San Francisco, 1973), p. 302.

2. B. Kostant: Lecture Notes in Mathematics, 170 (Springer, Berlin, 1970).

   J-M. Souriau: "Structure des Systèmes Dynamiques" (Dunod, Paris, 1970).

3. For example, see R. Abraham: Foundations of Mechanics (Benjamin, Reading, Mass., 1967) or A. Weinstein: Advances in Mathematics, 6, 329 (1971) for a noninductive proof valid in infinite dimensional manifolds.

4. It is sometimes clearer to make this distinction between coordinates on X and coordinates on TX or T^*X, though this will not always have been done in the following.

5. In a compact symplectic manifold (M,ω), ω cannot be exact, for if there existed a 1-form θ such that $\omega = d\theta$ then the natural volume element

$$\omega^n = \omega \wedge \omega \wedge \dots \wedge \omega$$

would also be exact: in fact

$$\omega^n = d(\omega \wedge \dots \omega \wedge \theta)$$

But then

$$\int_M \omega^n = 0$$

by Stokes's theorem (since M is compact). This is a contradiction.

6. S. Kobayashi and K. Nomizu: "Foundations of Differential Geometry", p.149 (Interscience, New York, 1969).

7. In local canonical coordinates, the integral curves of ξ_ϕ (for real $\phi \in C^\infty(M)$) are given by Hamilton's equations:

$$\dot{q}^a = \frac{\partial\phi}{\partial p_a} \quad \dot{p}_a = -\frac{\partial\phi}{\partial q^a}$$

8. The Lagrange bracket $\{\cdot,\cdot\}$ can also be defined in terms of ω : if $\xi,\zeta \in \mathcal{U}(M)$ then $\{\xi,\zeta\}$ is the C^∞ function:

$$\{\xi,\zeta\} = 2\omega(\xi,\zeta) \quad .$$

9. The operator $\xi \lrcorner : \Omega^m(M) \to \Omega^{m-1}(M)$ (where $\Omega^m(M)$ is the space of complex m-forms on M) is defined by:

$$\xi\lrcorner : \alpha \longrightarrow \xi\lrcorner\alpha = m \,.\, \alpha(\xi, \cdot \,, \ldots , \cdot); \; \alpha \in \Omega^m(M).$$

10. The symbol ξ_ϕ retains the special meaning assigned to it in §3.

11. See appendix C.

12. $[G,G]$ is the subspace of G spanned by all commutators.

13. B. Kostant: Op.Cit., p.133.

14. In local canonical coordinates, $\omega^n = d^n p \, d^n q$.

15. The substitution of $\pi^*(\omega)$ for $d\alpha$ is possible since ∇ has curvature ω.

16. There is an analogy in differential geometry: a vector field can be thought of either as a section of the tangent bundle or as a set of functions (components) on the bundle of frames.

17. The factor $-2\pi i$ is conventional: in the units used, $h = 1$.

18. Of course, there is the usual problem that, in general, δ_ϕ is not a well defined operator on the whole of H.

19. If the dynamics of the system are generated by a Hamiltonian h and if the system is degenerate (that is ξ_h has closed orbits) then prequantization leads directly to the Bohr-Sommerfeld condition on the energy levels: these are given by the solutions h_o of the eigenvalue equation for δ_h:

$$(\xi_h + 2\pi i\,(\xi_h \lrcorner\, \theta) - 2\pi i.h)\chi = -\,2\pi i.h_o\,\chi\,.$$

On the surface $h = h_o$, this equation takes the form:

$$\xi_h\chi + 2\pi i\,(\xi_h \lrcorner\, \theta)\chi = 0$$

from which it is clear that for χ to be single valued, the integral of θ around each orbit in the surface $h = h_o$ must be an integer. This is the Bohr-Sommerfeld condition. According to R. Blattner ("Quantization and Representation Theory", Proceeding's of Symposia in Pure Mathematics (A.M.S.), Vol 26, 1973. The inclusion of $^1/_2$-forms and the use of the full quantization scheme leads to the Maslov correction to the Bohr-Sommerfeld rule (see V.I. Arnol'd: Funct. Anal. and its Appl., 1, 1-13 (1967).

20. The integral of a density μ is defined locally (in coordinates $\{x^a\}$) by:

$$\int_U \mu = \int_U \mu(x, dx^1, \ldots, dx^n)\, d^n x \ ; \quad U \subset X$$

This is independent of the choice of coordinates by the transformation properties of μ. The integral is defined globally using a partition of unity.

21. That is, tangent to the fibres in E(M).

22. This can be shortened to: $\mathcal{L}_\eta P = 0$.

23. The $^1/_2$-density scheme is presented by Blattner, who also introduces the concept of a $^1/_2$-P-form: R. Blattner: Quantization and Representation Theory: in: A.M.S. Proc. of Symposia in Pure Math. 26 (1973). The principal motivation for the introduction of $^1/_2$-forms comes from Maslov's work on the asymptotic oscillatory solutions of partial differential equations: V.P. Maslov: Théorie des Perturbations et Méthodes Asymptotiques: Dunod-Gauthier-Villars, Paris (1972); see also: V.I. Arnol'd: Funct. Anal. and its Appl., 1, 1-13 (1967); J.J. Duistermaat: Commun. Pure and Appl. Math., 27, 207 (1974); L. Hörmander: Acta. Math., 127, 79 (1971). Essentially, the use of $^1/_2$-forms allows one to include within geometric quantization a phenomenon first described by Gouy;

namely that, in geometric optics limit (and also in the WKB approximation) the phase of the wave function jumps discontinuously at the caustics in the classical trajectories: L.G. Gouy: C.R. Acad. Sci. Paris 110, 1251 (1890). This phenomenon also leads to Maslov's correction to the Bohr-Sommerfeld condition (see note 19). A related motivation comes from Dixmier's work on solvable Lie algebras: the $^1/_2$-form concept is closely linked to his idea of a 'representation tordue': J. Dixmier: Algèbres Enveloppantes: Gauthier-Villars, Paris (1974). The way in which $^1/_2$-forms enable one to deal with caustics is indicated in §9, example 1.

24. $L \otimes L^P$ denotes the tensor product of L and L^P. See appendix A and note 49.

25. If $\xi, \zeta \in U_P(M)$ then $\nabla_\xi s = 0$ and $\nabla_\zeta s = 0$ implies $\nabla_{[\xi,\zeta]} s = 0$ since $\omega(\xi,\zeta) = 0$. Thus this condition is self consistent.

26. At this point one uses the condition that the integral surfaces of D are simply connected.

27. An alternative formulation of this condition is

$$[f, [g, \phi]] = 0$$

for all $f, g \in C^\infty(M)$ constant in the directions in P.

28. 'BKS' stands for 'Blattner, Kostant and Sternberg'. This construction is described in more detail by Blattner, who also describes the formal quantization of observables which do not preserve the polarization: R. Blattner: op. cit.

29. If P_1 and P_2 are not everywhere transverse, or if more than two polarizations are involved, a more complicated construction is needed: see §9, example 1.

30. I.M. Gel'fand and G.E. Shilov; "Generalized Functions", Vol. 1. (Academic Press, New York, 1964), p. 34.

31. This point is discussed further in appendix C.

32. Of course, what this does not do is tell one how these elementary classical systems are to be interpreted physically. However, considerable progress has been made on this question by Souriau (op. cit., Ch. III) and, more recently, by Penrose (in the relativistic context): R. Penrose: "Twistor Theory": in: Quantum Gravity: eds: C. Isham, R. Penrose and D. Sciama (Clarendon Press, Oxford, 1975).

33. Condition 2) is stronger than necessary: see Kostant, op. cit., p. 177.

34. Conversely, each orbit M_{f_o} is a Hamiltonian G-space: λ is defined by:

$$(\lambda(X))(f) = f(X) \; ; \quad f \in M_{f_o}, \quad X \in G \; .$$

35. If G is not simply connected then this gives a representation of the universal covering group. For example if $G = SO(1,3)$ then the construction gives representation of $SL(2,\mathbb{C})$ and thus leads naturally to the spinor concept.

36. These phase spaces are described by H.P. Künzle: J. Math. Phys. 13, 739 (1972).

37. For more information on this approach to the WKB approximation, see: J.J. Duistermaat, op. cit.; A. Voros: Semi-Classical Approximations (preprint, Saclay, 1974) and references therein. There is a possible cause for confusion in comparing the treatment given here with the discussion in Duistermaat's paper: when I define the local polarization generated by a family of functions S_k, I am implicitly assuming not only that there is just one surface $\phi_k(X)$ through each point of the region of phase space under consideration, but also that $\partial S/\partial k_a \neq 0$ in this region.

38. For an alternative treatment using a third polarization (which is neither real nor Kähler) see: D.J. Simms: "Metalinear Structures and the Quantization of the Harmonic Oscillator": in: International Colloquium on Sympletic Mechanics and Mathematical Physics (CNRS, Aix-en-Provence, 1974). Also, V.I. Bargmann: Commun. Pure Appl. Math., 14, 187 (1961).

39. V.I. Arnol'd: op. cit.

40. J.M. Souriau: op. cit., ch. V; P. Renouard: Thesis (Paris, 1969); A. Carey: Thesis (Oxford, 1975).

41. For an explanation of the notation, see: F.A.E. Pirani: "Introduction to Gravitational Radiation Theory": in: The Proceedings, Brandeis Summer Institute, 1964 (Prentice Hall, New Jersey, 1965); see also: R. Penrose: "The Structure of Space-Time": in: Battelle Rencontres: eds: C.M. DeWitt and J.A. Wheeler (Benjamin, New York, 1968).

42. R. Penrose: in: Quantum Gravity: eds: C. Isham, R. Penrose and D. Sciama (Clarendon Press, Oxford, 1975).

43. T is 'twistor space': see note 42.

44. Note the strong formal resemblance to the harmonic oscillator problem.

45. For more detail, see D.J. Simms: "Equivalence of Bohr-Sommerfeld Kostant-Souriau and Pauli quantization of the Kepler problem"; in: The Proceedings of the Colloquium on Group Theoretical Methods in Physics (CNRS, Marseille, 1972).

46. $E_{\chi,s}$ is simply connected.

47. R. Godement: "Theorie des Faisceaux" (Hermann, Paris); p. 213.

48. I. Vaisman: "Cohomology and Differential Forms" (Dekker, New York, 1973); p. 90.

49. The tensor product $L_1 \otimes L_2$ of two line bundles L_1 and L_2 (over M) is defined as follows: choose local systems $\{U_i, \underset{(1)}{s}{}_i\}$ and $\{U_i, \underset{(2)}{s}{}_i\}$ for L_1 and L_2. If the corresponding transition functions are:

$$\overset{(1)}{c}_{ij} : U_i \cap U_j \to \mathbb{C}^*$$

$$\overset{(2)}{c}_{ij} : U_i \cap U_j \to \mathbb{C}^*$$

then the transition functions for $L_1 \otimes L_2$ are:

$$c_{ij} = \overset{(1)}{c}_{ij} \cdot \overset{(2)}{c}_{ij} : U_i \cap U_j \to \mathbb{C}^*$$

50. For more details of this proof, see Kostant, op. cit., p. 133.

51. That is: $hf = f \; \forall \; f \in F$ if and only if h is the identity in G.

52. In the following, the term fibre bundle will be used as an abbreviation for "locally trivial fibre bundle with structure group".

53. For more information on Lie algebra cohomology see: C. Chevalley and S. Eilenberg: Trans Am. Math. Soc., 63, 85 (1948); D.J. Simms: "Projective Representations, Symplectic Manifolds and Extensions of Lie Algebras" Mimeographed lecture notes (CNRS, Marseille, 1969)
R. Hermann: "Vector Bundles in Mathematical Physics" (Benjamin, New York, 1970).

54. D.J. Simms: "Projective Representations, Symplectic Manifolds and Extensions of Lie Algebras": Mimeographed lecture notes (CNRS, Marseille, 1969).

55. R. Hermann, op.cit., Vol II, p.78.

56. It is an immediate consequence of eqn. C.5 that $L = [L,L]$ (for any semi-simple L). For suppose that $L \neq [L,L]$. Choose $Z \neq 0 \in [L,L]^{\perp}$ (orthogonal complement with respect to $\langle \cdot , \cdot \rangle$). Then, for any $X,Y \in L$:

$$0 = \langle [X,Y],Z \rangle = \langle X,[Y,Z] \rangle \quad .$$

whence:

$$[Y,Z] = 0 \qquad \forall \quad Y \in L,$$

that is, $\mathrm{ad}\, Z = 0$. But then

$$\langle Z,X\rangle = 0 \quad \forall\ X \in L$$

implying $Z = 0$, a contradiction.

57. For example, $C^{\infty}_{\mathbb{R}}(M)$ is a central extension of $\mathcal{A}(M)$.

58. As a Lie algebra, E is the semi-direct sum of $\iota(L)$ and $\mathbb{R}$, in the sense that if $\ell \in \iota(L) \subset E$ and if $r \in \mathbb{R} \subset E$ then $[r,\ell] \in \mathbb{R} \subset E$.

Since these notes were written, a number of new developments have taken place; some of these are described in:

1. Contributions by B. Kostant, E. Onofri, D.J. Simms, J. Sniatycki, J-M. Souriau, J.A. Wolf and N.M.J. Woodhouse in: The Proceedings of the Colloquium on Group Theoretical Methods in Physics, Nijmegen, 1975.
2. Contributions by R.J. Blattner, J. Ehlers, K. Gawedzki, B. Kostant, A. Lichnerowicz, E. Onofri, D.J. Simms, J. Sniatycki, J-M. Souriau and S. Sternberg in: The proceedings of the Conference on Differential Geometrical Methods in Mathematical Physics, Bonn 1975.
3. B. Kostant: "On the Definition of Quantization" (Preprint, MIT, 1975).
4. J.H. Rawnsley: "Diagonal Quantization of Hamiltonians with Periodic Flows" (Preprint, Mathematisches Institut, Bonn, 1975).

Selected Issues from

Lecture Notes in Mathematics

Vol. 507: M. C. Reed, Abstract Non-Linear Wave Equations. VI, 128 pages. 1976.

Vol. 501: Spline Functions, Karlsruhe 1975. Proceedings. Edited by K. Böhmer, G. Meinardus, and W. Schempp. VI, 421 pages. 1976.

Vol. 495: A. Kerber, Representations of Permutation Groups II. V, 175 pages. 1975.

Vol. 490: The Geometry of Metric and Linear Spaces. Proceedings 1974. Edited by L. M. Kelly. X, 244 pages. 1975.

Vol. 489: J. Bair and R. Fourneau, Etude Géométrique des Espaces Vectoriels. Une Introduction. VII, 185 pages. 1975.

Vol. 485: J. Diestel, Geometry of Banach Spaces – Selected Topics. XI, 282 pages. 1975.

Vol. 484: Differential Topology and Geometry. Proceedings 1974. Edited by G. P. Joubert, R. P. Moussu, and R. H. Roussarie. IX, 287 pages. 1975.

Vol. 481: M. de Guzmán, Differentiation of Integrals in R^n. XII, 226 pages. 1975.

Vol. 480: X. M. Fernique, J. P. Conze et J. Gani, Ecole d'Eté de Probabilités de Saint-Flour IV–1974. Edité par P.-L. Hennequin. XI, 293 pages. 1975.

Vol. 477: Optimization and Optimal Control. Proceedings 1974. Edited by R. Bulirsch, W. Oettli, and J. Stoer. VII, 294 pages. 1975.

Vol. 474: Séminaire Pierre Lelong (Analyse) Année 1973/74. Edité par P. Lelong. VI, 182 pages. 1975.

Vol. 470: R. Bowen, Equilibrium States and the Ergodic Theory of Anosov Diffeomorphisms. III, 108 pages. 1975.

Vol. 468: Dynamical Systems – Warwick 1974. Proceedings 1973/74. Edited by A. Manning. X, 405 pages. 1975.

Vol. 464: C. Rockland, Hypoellipticity and Eigenvalue Asymptotics. III, 171 pages. 1975.

Vol. 463: H.-H. Kuo, Gaussian Measures in Banach Spaces. VI, 224 pages. 1975.

Vol. 461: Computational Mechanics. Proceedings 1974. Edited by J. T. Oden. VII, 328 pages. 1975.

Vol. 459: Fourier Integral Operators and Partial Differential Equations. Proceedings 1974. Edited by J. Chazarain. VI, 372 pages. 1975.

Vol. 458: P. Walters, Ergodic Theory – Introductory Lectures. VI, 198 pages. 1975.

Vol. 449: Hyperfunctions and Theoretical Physics. Proceedings 1973. Edited by F. Pham. IV, 218 pages. 1975.

Vol. 448: Spectral Theory and Differential Equations. Proceedings 1974. Edited by W. N. Everitt. XII, 321 pages. 1975.

Vol. 447: S. Toledo, Tableau Systems for First Order Number Theory and Certain Higher Order Theories. III, 339 pages. 1975.

Vol. 446: Partial Differential Equations and Related Topics. Proceedings 1974. Edited by J. A. Goldstein. IV, 389 pages. 1975.

Vol. 445: Model Theory and Topoi. Edited by F. W. Lawvere, C. Maurer, and G. C. Wraith. III, 354 pages. 1975.

Vol. 444: F. van Oystaeyen, Prime Spectra in Non-Commutative Algebra. V, 128 pages. 1975.

Vol. 443: M. Lazard, Commutative Formal Groups. II, 236 pages. 1975.

Vol. 442: C. H. Wilcox, Scattering Theory for the d'Alembert Equation in Exterior Domains. III, 184 pages. 1975.

Vol. 441: N. Jacobson, PI-Algebras. An Introduction. V, 115 pages. 1975.

Vol. 440: R. K. Getoor, Markov Processes: Ray Processes and Right Processes. V, 118 pages. 1975.

Vol. 439: K. Ueno, Classification Theory of Algebraic Varieties and Compact Complex Spaces. XIX, 278 pages. 1975.

Vol. 438: Geometric Topology. Proceedings 1974. Edited by L. C. Glaser and T. B. Rushing. X, 459 pages. 1975.

Vol. 437: D. W. Masser, Elliptic Functions and Transcendence. XIV, 143 pages. 1975.

Vol. 436: L. Auslander and R. Tolimieri, Abelian Harmonic Analysis, Theta Functions and Function Algebras on a Nilmanifold. V, 99 pages. 1975.

Vol. 435: C. F. Dunkl and D. E. Ramirez, Representations of Commutative Semitopological Semigroups. VI, 181 pages. 1975.

Vol. 434: P. Brenner, V. Thomée, and L. B. Wahlbin, Besov Spaces and Applications to Difference Methods for Initial Value Problems. II, 154 pages. 1975.

Vol. 433: W. G. Faris, Self-Adjoint Operators. VII, 115 pages. 1975.

Vol. 432: R. P. Pflug, Holomorphiegebiete, pseudokonvexe Gebiete und das Levi-Problem. VI, 210 Seiten. 1975.

Vol. 431: Séminaire Bourbaki – vol. 1973/74. Exposés 436–452. IV, 347 pages. 1975.

Vol. 430: Constructive and Computational Methods for Differential and Integral Equations. Proceedings 1974. Edited by D. L. Colton and R. P. Gilbert. VII, 476 pages. 1974.

Vol. 429: L. Cohn, Analytic Theory of the Harish-Chandra C-Function. III, 154 pages. 1974.

Vol. 428: Algebraic and Geometrical Methods in Topology, Proceedings 1973. Edited by L. F. McAuley. XI, 280 pages. 1974.

Vol. 427: H. Omori, Infinite Dimensional Lie Transformation Groups. XII, 149 pages. 1974.

Vol. 426: M. L. Silverstein, Symmetric Markov Processes. X, 287 pages. 1974.